SpringerBriefs in Optimization

Series Editors
Sergiy Butenko
Mirjam Dür
Panos M. Pardalos
János D. Pintér
Stephen M. Robinson
Tamás Terlaky
My T. Thai

SpringerBriefs in Optimization showcases algorithmic and theoretical techniques, case studies, and applications within the broad-based field of optimization. Manuscripts related to the ever-growing applications of optimization in applied mathematics, engineering, medicine, economics, and other applied sciences are encouraged.

More information about this series at http://www.springer.com/series/8918

Adam B. Levy

Attraction in Numerical Minimization

Iteration Mappings, Attractors, and Basins of Attraction

 Springer

Adam B. Levy
Bowdoin College
Brunswick
ME, USA

ISSN 2190-8354   ISSN 2191-575X   (electronic)
SpringerBriefs in Optimization
ISBN 978-3-030-04048-2   ISBN 978-3-030-04049-9   (eBook)
https://doi.org/10.1007/978-3-030-04049-9

Library of Congress Control Number: 2018963116

Mathematics Subject Classification: 49J53, 90C, 49M, 65K

This Springer imprint is published by the registered company Springer Nature Switzerland AG
The registered company address is: Gewerbestrasse 11, 6330 Cham, Switzerland

To Sarah

Preface

Equilibria are fundamental objects of study in dynamical systems theory because they can be used to classify and understand the behavior of the system. We introduce a parallel concept of "attractors" in numerical minimization and use various related objects from dynamical systems theory as inspiration for analogues in numerical minimization. We then employ our new tools to carefully analyze a variety of particular examples in numerical minimization in order to demonstrate how these tools can enrich our understanding. Our definition of attractor rests on the platform of "iteration mappings" which are a special type of "multiset-mapping" that we also introduce and develop here.

Set-valued mappings associate sets to individual inputs, generalizing the notion of a single-valued mapping. Set-valued mappings arise naturally in many areas, for instance, in optimization where solutions may not be unique and where generalized (and set-valued) notions of continuity and derivatives allow the variational analysis of problems with constraints and/or non-smooth objective functions. The origins of variational analysis can be traced to at least 1925 in the work of Vasilesco [24], and this research area remains very active today. Some important books covering various stages of this subject to date include [4, 6, 12, 18, 21].

Numerical variational analysis is a related area that uses generalized notions of continuity and differentiability to analyze the convergence and stability of numerical optimization methods, and the focus of this analysis has traditionally been on individual "iterate" vectors generated by a method at each step. It has long been recognized (e.g., [25]) that we may conceive of these iterates as the input and/or output of a set-valued iteration mapping because the next iterate may need to be chosen from several promising trial vectors generated by the current iterate. This formulation is important since so many numerical optimization methods use multiple trial vectors to determine an iterate vector at each step (e.g., pattern-search methods, dating back at least to [11], or any method employing a backtracking line search). However, this formulation is limited to methods that use a single iterate vector to generate the next trial vectors, and some numerical optimization methods actually generate and use multiple iterate vectors at each step, including the Nelder-Mead method [19] and trust-region methods using interpolating model functions (e.g., [8]). Therefore, to more thoroughly represent the evolution of optimization methods from one iteration to the next via an iteration mapping, we will extend the notion of set-valued mapping to accommodate set inputs as well as set outputs. Moreover, the vectors generated at each iteration of an optimization method sometimes appear with copies (especially in the limit), so the input and output of an iteration mapping actually might be "multisets." Because of this, we will define and develop multiset-mappings that take multisets to multisets; and we call the special multisets generated by iteration mappings "iterate-multisets."

In order to study the convergence properties of the sequences of iterate-multisets generated by numerical minimization methods, we introduce a generalized continuity property for multiset-mappings called "calmness." A localized version of calmness for set-valued mappings was introduced in [20], and the version we introduce here for multiset-mappings uses pre-distance functions on the domain and range spaces. We characterize calmness for multiset-mappings via a generalized derivative for multiset-mappings that we also introduce here, and we use calmness to provide two versions of fixed-point theorems for iteration mappings that generalize the Banach fixed-point theorem [5, Theorem 6]. A crucial concept for iteration mappings is the "viability" of initial iterate-multisets, which identifies the iterate-multisets from which the repeated application of the iteration mapping produces a sequence of non-empty iterate-multisets. We precisely define iteration mappings for three different well-known minimization methods: coordinate-search, steepest-descent (with three different line-searches), and Nelder-Mead; and we discuss viability in each case.

We define appropriate notions of stability and asymptotic stability for attractors in numerical optimization, where the second notion involves the additional presence of a positive "radius of attraction" (signaling that the attractor attracts every viable initial iterate-multiset whose elements are close to the attractor). We apply our generalized Banach fixed-point theorems to deduce conditions on the iteration mapping that ensure a positive radius of attraction. We also give conditions under which a positive radius of attraction implies that the attractor is a local minimizer, and we provide a companion result involving a weaker notion of "radius of restricted attraction." In addition, we prove that the reverse implication of the companion result holds when the attractor is stable. These results rely on important properties of "local dense viability" (viable initial iterate-multisets are sufficiently abundant near an attractor) and "minvalue-monotonicity" (the minimum value of the objective function does not increase with iteration). All of these notions depend on both the objective function being minimized and the method of numerical minimization. So, for instance, an attractor for one method might not be an attractor for another, even if the objective function remains the same. This is fundamentally different from the objects from dynamical systems we use for inspiration, since those are fixed by the system and do not vary with the method one might use to explore the system.

Another object we adapt from dynamical systems is the "basin of attraction" associated with an attractor in numerical optimization; and we pair this with companion notions of "basin size" and "basin entropy." Both of these notions are computed by simulation, and the latter quantifies the complexity of basin boundaries. We illustrate simulated basins of attraction for four example objective functions using the same four different numerical methods for each (coordinate-search and steepest-descent with three different line-searches). For all 16 combinations, we compute and classify basin size, as well as computing and illustrating basin entropy. We apply the Nelder-Mead method to the four example functions in a separate section because its non-singleton iterate-multisets require different illustrations than the other methods.

Using the tools we have developed here, we investigate the practical significance of two well-known counterexamples to good convergence behavior in numerical minimization: the canoe function with coordinate-search and McKinnon's function [17] with Nelder-Mead. We use our notions of basin size and basin entropy to quantify the extent to which initial data are likely to lead to undesirable consequences. This investigation was stimulated by the curious fact that the Nelder-Mead method is widely used despite its possibility of failure for high-dimensional problems [23], as well as the impossibility of traditional convergence theory to support it (as demonstrated by McKinnon [17]). The authors of [15] suggest some possible reasons for the continued popularity of the Nelder-Mead method, and one minor consequence of our work here is to add to their list by demonstrating that the theory-stifling result of [17] can be viewed as practically insignificant.

Harpswell, ME, USA

Adam B. Levy

December 2017

Contents

Chapter 1
Multisets and Multiset Mappings

For the purposes of the investigation here, *multisets* are collections X of elements x from a normed vector space $\mathscr{X}$ where the same elements may appear in multiple instances (e.g., so $\{1, 1\}$ is different from $\{1\}$). (For example, see [13, Notes, page 636] for a discussion on the terminology of multisets.) In general, the order of elements in a multiset is not significant (e.g., so $\{1, 2\}$ is the same as $\{2, 1\}$), though we will make use of ordered multisets where order does matter. In that case, we will sometimes make use of the *first* mapping which returns the element First(X) appearing first in the multiset X. For example:

$$\text{First}\,(\{1, 2\}) = 1 \quad \text{and} \quad \text{First}\,(\{2, 1\}) = 2.$$

The *multiplicity function* $\delta_X : \mathscr{X} \to \mathbb{N}$ is used to count the instances in a multiset X of each element $x \in \mathscr{X}$. For example, $\delta_{\{1,1,2\}}(0) = 0$, $\delta_{\{1,1,2\}}(1) = 2$, and $\delta_{\{1,1,2\}}(2) = 1$. Evidently, any multiset is determined entirely by its multiplicity function, and when X is a standard set (without repeated elements), the multiplicity function reduces to the *indicator function*; returning 1 for elements in the set, and 0 otherwise. The *cardinality* $|X|$ of a multiset X is the number of elements in it (counting multiplicity). The usual set-operations of containment and intersection are extended to multisets A and B via the multiplicity function as follows:

$$A \subseteq B \iff \delta_A(x) \le \delta_B(x) \quad \forall x \in \mathscr{X}$$

$$\delta_{A \cap B}(x) = \min\{\delta_A(x), \delta_B(x)\} \quad \forall x \in \mathscr{X}.$$

In many cases, the multiplicity of the elements in a multiset will not be as significant as the fact that the elements appear in the multiset at all. For this reason, we define the *filter* of a multiset X to be the underlying set Filter(X) of elements appearing in X. Element inclusion $x \in X$ for a multiset X is defined by the usual element inclusion in the set Filter(X):

A. B. Levy, *Attraction in Numerical Minimization*, SpringerBriefs in Optimization,
https://doi.org/10.1007/978-3-030-04049-9_1

$$x \in X \iff x \in \text{Filter}(X) \iff \delta_{\text{Filter}(X)}(x) = 1 \iff \delta_X(x) > 0.$$

Standard set notation has the consequence that $\{x \in X\}$ is the filter of the multiset X, and not necessarily the multiset X itself. To rectify this, we introduce sub-delta notation $x \in_\delta X$ when we mean to use all $\delta_X(x)$ copies of each element $x \in X$. Thus, we have the two identities:

$$\text{Filter}(X) = \{x \in X\} \text{ and } X = \{x \in_\delta X\}.$$

In order to study sequences (and subsequences) of multisets, we will use the notation:

$$\mathcal{N}_\infty^\# := \{N \subseteq \mathbb{N} : N \text{ infinite}\}$$

to denote the infinite subsets of $\mathbb{N}$ (e.g., see [21]). For a sequence $\{X^k\}$ of multisets, the *outer limit* is the set:

$$\limsup_{k \to \infty} X^k := \{x^\infty : \exists N \in \mathcal{N}_\infty^\#, \exists x^k \in X^k \ (k \in N) \text{ with } x^k \xrightarrow{N} x^\infty\},$$

which collects the cluster points of sequences of elements $x^k \in X^k$ when the multisets X^k are all non-empty. Note that the outer limit can be empty (e.g., if all of the multisets X^k are empty for sufficiently large k).

A *multiset mapping* $S : \mathscr{X} \rightrightarrows \mathscr{Y}$ between two normed vector spaces $\mathscr{X}$ and $\mathscr{Y}$ maps multisets X of elements from $\mathscr{X}$ to multisets $S(X)$ of elements from $\mathscr{Y}$. This construction generalizes set-valued mappings (also known as multifunctions) which take single vectors $x \in \mathscr{X}$ into sets $S(x) \subseteq \mathscr{Y}$, and which themselves generalize single-valued mappings. Just like any other mapping, a multiset mapping S may only generate (non-empty) output for some inputs. The *domain* of a multiset mapping S is the collection of multisets X for which $S(X)$ is non-empty:

$$\text{dom}(S) := \{X : S(X) \neq \emptyset\},$$

and the *range* is the set of elements appearing in some image-multiset:

$$\text{rge}(S) := \{y : \exists X \text{ with } y \in S(X)\}.$$

1.1 Pre-distance Functions

The continuity property we will introduce for multiset mappings depends on very general extended real-valued *pre-distance* functions **d** operating on pairs of multisets (A, B) from the same normed vector space. The only requirements on the pre-distance functions **d** are as follows:

(i) Non-negative on non-empty multisets:

$$\mathbf{d}(A, B) \geq 0 \text{ when } A \neq \emptyset \text{ and } B \neq \emptyset,$$

(ii) Equal the usual distance when applied to single vectors a and b:

$$\mathbf{d}(a, b) = \|a - b\|.$$

We will add subscripts to the generic notation $\mathbf{d}$ in order to distinguish one particular pre-distance function from another. For instance, some useful examples for this investigation are

$$\mathbf{d}_{\text{inf}}(A, B) = \inf_{a \in A} \inf_{b \in B} \|a - b\|$$

$$\mathbf{d}_{\text{exc}}(A, B) = \sup_{a \in A} \inf_{b \in B} \|a - b\|$$

$$\mathbf{d}_{\text{sup}}(A, B) = \sup_{a \in A} \sup_{b \in B} \|a - b\|.$$

The subscript on the pre-distance $\mathbf{d}_{\text{exc}}$ recognizes that this function is known as the "excess" of A over B.

Recall that a genuine distance function must be positive-definite: $\mathbf{d}(A, B) = 0$ if and only if $A = B$, and notice that the three pre-distances above all fail this condition:

$$\mathbf{d}_{\text{inf}} (A, B) = 0 \iff A \cap B \neq \emptyset$$

$$\mathbf{d}_{\text{exc}} (A, B) = 0 \iff A \neq \emptyset \text{ and } A \subseteq B \tag{1.1}$$

$$\mathbf{d}_{\text{sup}} (A, B) = 0 \iff A = B \text{ is a singleton.}$$

A genuine distance function must also be symmetric $\mathbf{d}(A, B) = \mathbf{d}(B, A)$, and the excess function $\mathbf{d}_{\text{exc}}$ doesn't even have this property.

The obvious hierarchy among the three pre-distance functions above will be useful in the sequel:

$$\mathbf{d}_{\text{inf}}(A, B) \leq \mathbf{d}_{\text{exc}}(A, B) \leq \mathbf{d}_{\text{sup}}(A, B) \quad \text{for all } A \neq \emptyset \text{ and } B \neq \emptyset. \tag{1.2}$$

Note that these inequalities may be strict. For example, consider the simple case of the finite sets $A = \{0, 1\}$ and $B = \{0\}$. We have $\mathbf{d}_{\text{inf}}(A, B) = 0 < 1 = \mathbf{d}_{\text{exc}}(A, B)$. On the other hand, we have $\mathbf{d}_{\text{exc}}(B, A) = 0 < 1 = \mathbf{d}_{\text{sup}}(B, A)$. This example also illustrates the lack of symmetry in $\mathbf{d}_{\text{exc}}$.

The following lemma provides a useful geometric formulation of $\mathbf{d}_{\text{exc}}$ in terms of the unit ball about the zero-vector in the normed vector space.

Lemma 1.1 *When $A \neq \emptyset$, $\mathbf{d}_{\mathrm{exc}}(A, B) = \inf\{\eta \geq 0 : A \subseteq B + \eta\,\mathbb{B}\}$.*

Proof In the case when $B = \emptyset$, we know $\mathbf{d}_{\mathrm{exc}}(A, B) = \infty$ and $\inf\{\eta \geq 0 : A \subseteq B + \eta\,\mathbb{B}\} = \infty$; so, we assume $B \neq \emptyset$.

To prove $\mathbf{d}_{\mathrm{exc}}(A, B) \leq \inf\{\eta \geq 0 : A \subseteq B + \eta\,\mathbb{B}\}$, we notice that if $\eta \geq 0$ is such that $A \subseteq B + \eta\,\mathbb{B}$, then for any vector $a \in A$, there exist $b \in B$ and $e \in \mathbb{B}$ such that $a = b + \eta\,e$. It follows immediately that $\|a - b\| = \eta\,\|e\| \leq \eta$, from which we deduce that $\mathbf{d}_{\mathrm{exc}}(A, B) \leq \eta$. The inequality then follows since this is true for any such $\eta \geq 0$.

We establish the opposite inequality by setting $\bar{\eta} = \mathbf{d}_{\mathrm{exc}}(A, B)$ and recalling from the definition of $\mathbf{d}_{\mathrm{exc}}(A, B)$ that this means for every vector $a \in A$, there exists $b_a \in B$ with $\|a - b_a\| \leq \bar{\eta}$. We conclude in particular that

$$\{a\} \subseteq \{b_a\} + \bar{\eta}\,\mathbb{B} \subseteq B + \bar{\eta}\,\mathbb{B},$$

and since this holds for every $a \in A$, we conclude that $A \subseteq B + \bar{\eta}\,\mathbb{B}$. Thus, $\bar{\eta} \geq \inf\{\eta \geq 0 : A \subseteq B + \eta\,\mathbb{B}\}$ and we are done since $\bar{\eta} = \mathbf{d}_{\mathrm{exc}}(A, B)$.   □

Other pre-distance functions of interest include

$$\mathbf{d}_{\mathrm{PH}}(A, B) = \sup_{c}\left|\inf_{a \in A}\|a - c\| - \inf_{b \in B}\|b - c\|\right|$$

$$\mathbf{d}_{\mathrm{PHB}}(A, B) = \inf\{\eta \geq 0 : A \subseteq B + \eta\,\mathbb{B}, \; B \subseteq A + \eta\,\mathbb{B}\},$$

where the PH in the subscript honors Pompeiu and Hausdorff. For non-empty closed sets A and B in $\mathbb{R}^n$, these represent the same (genuine) distance function (see [21, Example 4.13]). However, neither of these is positive-definite for multisets:

$$\text{e.g., } \mathbf{d}_{\mathrm{PH}}(\{1, 1\}, \{1\}) = \mathbf{d}_{\mathrm{PHB}}(\{1, 1\}, \{1\}) = 0.$$

Related functions from [21] are defined in terms of $\rho \in [0, \infty)$ as follows:

$$\mathbf{d}_{\rho}(A, B) = \sup_{\|c\| \leq \rho}\left|\inf_{a \in A}\|a - c\| - \inf_{b \in B}\|b - c\|\right|$$

$$\mathbf{d}_{\rho\mathbb{B}}(A, B) = \inf\{\eta \geq 0 : A \cap \rho\,\mathbb{B} \subseteq B + \eta\,\mathbb{B}, \; B \cap \rho\,\mathbb{B} \subseteq A + \eta\,\mathbb{B}\}.$$

However, neither of these is a legitimate pre-distance function here since the condition (ii) $\mathbf{d}(a, b) = \|a - b\|$ is violated when $\|a\| > \rho$ and $\|b\| > \rho$.

1.2   Calmness

Recall (e.g., [21]) that a single-valued mapping $S : \mathbb{R}^m \to \mathbb{R}^n$ is *calm* at $\bar{x} \in \mathrm{dom}(S)$ with radius $\delta > 0$ and modulus $L \in (0, \infty)$ if

$$\|S(x) - S(\bar{x})\| \leq L\,\|x - \bar{x}\| \text{ when } x \in \mathrm{dom}(S) \text{ and } \|x - \bar{x}\| \leq \delta.$$

Note that this continuity property is different from (and weaker than) a typical Lipschitz continuity property since the comparison is only against the base vector $\bar{x}$. For any pre-distance functions $\mathbf{d}_{\mathscr{X}}$ (operating on multisets from $\mathscr{X}$) and $\mathbf{d}_{\mathscr{Y}}$ (operating on multisets from $\mathscr{Y}$), we say that a multiset mapping $S : \mathscr{X} \rightrightarrows \mathscr{Y}$ is $\frac{\mathbf{d}_{\mathscr{Y}}}{\mathbf{d}_{\mathscr{X}}}$-*calm at* $\bar{X} \in \mathrm{dom}(S)$ if there exist $L \in (0, \infty)$ and $\delta > 0$ such that

$$\mathbf{d}_{\mathscr{Y}}\left(S(X), S(\bar{X})\right) \le L\, \mathbf{d}_{\mathscr{X}}\left(X, \bar{X}\right) \text{ when } X \in \mathrm{dom}(S) \text{ and } \mathbf{d}_{\mathscr{X}}\left(X, \bar{X}\right) \le \delta. \tag{1.3}$$

In this case, we refer to L as a *calmness modulus* and to δ as a *calmness radius*. The notation $\frac{\mathbf{d}_{\mathscr{Y}}}{\mathbf{d}_{\mathscr{X}}}$ is motivated by the fact that the calmness inequality in (1.3) can be rewritten as:

$$\frac{\mathbf{d}_{\mathscr{Y}}\left(S(X), S(\bar{X})\right)}{\mathbf{d}_{\mathscr{X}}\left(X, \bar{X}\right)} \le L, \tag{1.4}$$

as long as the denominator is not zero. When the particular pre-distance functions are both the same type $\mathbf{d}$ and there is no ambiguity, we say simply that S is $\mathbf{d}$-*calm at* $\bar{X}$.

The following implications result immediately from the hierarchy of pre-distance functions (1.2).

Proposition 1.1 *For any pre-distance functions* $\mathbf{d}_{\mathscr{X}}$ *and* $\mathbf{d}_{\mathscr{Y}}$*, the implications hold that*

$$S \text{ is } \frac{\mathbf{d}_{\mathscr{Y}}}{\mathbf{d}_{\mathrm{inf}}}\text{-calm at } \bar{X} \implies S \text{ is } \frac{\mathbf{d}_{\mathscr{Y}}}{\mathbf{d}_{\mathrm{exc}}}\text{-calm at } \bar{X} \implies S \text{ is } \frac{\mathbf{d}_{\mathscr{Y}}}{\mathbf{d}_{\mathrm{sup}}}\text{-calm at } \bar{X}$$

$$S \text{ is } \frac{\mathbf{d}_{\mathrm{sup}}}{\mathbf{d}_{\mathscr{X}}}\text{-calm at } \bar{X} \implies S \text{ is } \frac{\mathbf{d}_{\mathrm{exc}}}{\mathbf{d}_{\mathscr{X}}}\text{-calm at } \bar{X} \implies S \text{ is } \frac{\mathbf{d}_{\mathrm{inf}}}{\mathbf{d}_{\mathscr{X}}}\text{-calm at } \bar{X}$$

Remark 1.1 None of the converse implications in Proposition 1.1 hold, as we show in the following four examples:

Counterexamples to Converse Implications

- $\frac{\mathbf{d}_{\mathscr{Y}}}{\mathbf{d}_{\mathrm{exc}}}$-calm at $\bar{X} \;\not\Longrightarrow\; S$ is $\frac{\mathbf{d}_{\mathscr{Y}}}{\mathbf{d}_{\mathrm{inf}}}$-calm at $\bar{X}$:

    Choose $\bar{X} = \{0\}$, $\mathbf{d}_{\mathscr{Y}} = \mathbf{d}_{\mathrm{exc}}$, and $S(X) = X$ with $\mathrm{dom}(S) = \{\{0\}, \{0, 1\}\}$. Then, we have

$$\mathbf{d}_{\mathrm{exc}}\left(S(\{0, 1\}), S(\bar{X})\right) = \mathbf{d}_{\mathrm{exc}}(\{0, 1\}, \{0\}) = 1;$$

    so S is $\mathbf{d}_{\mathrm{exc}}$-calm at $\bar{X}$ with calmness modulus $L = 1$ and radius $\delta = 1$. However, $L\, \mathbf{d}_{\mathrm{inf}}(\{0, 1\}, \bar{X}) = 0$ for any fixed $L \in (0, \infty)$.

(continued)

- $\frac{\mathbf{d}_{\mathscr{Y}}}{\mathbf{d}_{\sup}}$-calm at $\bar{X}$ $\not\Longrightarrow$ S is $\frac{\mathbf{d}_{\mathscr{Y}}}{\mathbf{d}_{\mathrm{exc}}}$-calm at $\bar{X}$

  Choose $\bar{X} = \{0, 1\}$, $\mathbf{d}_{\mathscr{Y}} = \mathbf{d}_{\sup}$, and $S(X) = X$ with $\mathrm{dom}(S) = \{\{0\}, \{0, 1\}\}$. Then, we have

$$\mathbf{d}_{\sup}\left(S(\{0\}), S(\bar{X})\right) = \mathbf{d}_{\sup}(\{0\}, \{0, 1\}) = 1;$$

  so S is $\mathbf{d}_{\sup}$-calm at $\bar{X}$ with calmness modulus $L = 1$ and radius $\delta = 1$. However, $L\,\mathbf{d}_{\mathrm{exc}}(\{0\}, \bar{X}) = 0$ for any fixed $L \in (0, \infty)$.

- $\frac{\mathbf{d}_{\mathrm{exc}}}{\mathbf{d}_{\mathscr{X}}}$-calm at $\bar{X}$ $\not\Longrightarrow$ S is $\frac{\mathbf{d}_{\sup}}{\mathbf{d}_{\mathscr{X}}}$-calm at $\bar{X}$:

  Choose $\bar{x} = 0$, $\mathbf{d}_{\mathscr{X}}(a, b) = |a - b|$, and $S(x) = \{0, 1\}$ with $\mathrm{dom}(S) = \mathbb{R}$. Then for any $x \in \mathbb{R}$, we have

$$\mathbf{d}_{\mathrm{exc}}\left(S(x), S(\bar{x})\right) = \mathbf{d}_{\mathrm{exc}}\left(\{0, 1\}, \{0, 1\}\right) = 0;$$

  so S is trivially $\frac{\mathbf{d}_{\mathrm{exc}}}{\mathbf{d}_{\mathscr{X}}}$-calm at $\bar{x}$. On the other hand, $\mathbf{d}_{\sup}\left(S(x), S(\bar{x})\right) = 1$ for any $x \in \mathbb{R}$, which clearly cannot be bounded above by $L\,|x|$ with a fixed $L \in (0, \infty)$ for all x near 0.

- $\frac{\mathbf{d}_{\inf}}{\mathbf{d}_{\mathscr{X}}}$-calm at $\bar{X}$ $\not\Longrightarrow$ S is $\frac{\mathbf{d}_{\mathrm{exc}}}{\mathbf{d}_{\mathscr{X}}}$-calm at $\bar{X}$:

  Choose $\bar{x} = 0$, $\mathbf{d}_{\mathscr{X}}(a, b) = |a - b|$, and $S(x) = \left\{\sqrt{|x|}, x\right\}$ with $\mathrm{dom}(S) = \mathbb{R}$. Then for any x less than or equal to 1 in absolute value, we have $\mathbf{d}_{\inf}\left(S(x), S(\bar{x})\right) = |x|$ which always equals $\mathbf{d}_{\mathscr{X}}(x, \bar{x}) = |x|$. Thus, S is $\frac{\mathbf{d}_{\inf}}{\mathbf{d}_{\mathscr{X}}}$-calm at $\bar{x}$ with calmness modulus $L = 1$ and radius $\delta = 1$. However, we also have

$$\mathbf{d}_{\mathrm{exc}}\left(S(x), S(\bar{x})\right) = \mathbf{d}_{\mathrm{exc}}\left(\left\{\sqrt{|x|}, x\right\}, \{0, 0\}\right) = \max\left\{\sqrt{|x|}, |x|\right\};$$

  which evaluates to $\sqrt{|x|}$ when $|x| \leq 1$. Since there is no fixed value of $L \in (0, \infty)$ for which $\sqrt{|x|} \leq L\,|x|$ for all x near 0, we conclude that S is not $\frac{\mathbf{d}_{\mathrm{exc}}}{\mathbf{d}_{\mathscr{X}}}$-calm at $\bar{x}$.

1.2.1   Calmness for Set-Valued Mappings

$\frac{\mathbf{d}_{\mathscr{Y}}}{\mathbf{d}_{\mathscr{X}}}$-calmness for multiset mappings not only generalizes calmness for single-valued mappings but also generalizes the following standard notion of a set-valued mapping S being *calm* at $\bar{x} \in \mathrm{dom}(S)$:

$$\exists L \in (0, \infty) \text{ and } \delta > 0 \text{ with } S(x) \subseteq S(\bar{x}) + L\,\|x - \bar{x}\|\,\mathbb{B} \text{ when } \|x - \bar{x}\| \leq \delta. \tag{1.5}$$

(For example, see [21, 9(30)] for this definition in the finite-dimensional case.)

Proposition 1.2 *A set-valued mapping S is calm at $\bar{x}$ if and only if S is $\frac{\mathbf{d}_{\mathrm{exc}}}{\mathbf{d}_{\mathscr{X}}}$-calm at $\bar{x}$ for every pre-distance function $\mathbf{d}_{\mathscr{X}}$.*

Proof ($\Rightarrow$) Since every pre-distance function $\mathbf{d}_{\mathscr{X}}$ satisfies $\mathbf{d}_{\mathscr{X}}(x, \bar{x}) = \|x - \bar{x}\|$ for single vectors, we can translate the assumption that S is calm at $\bar{x}$ (1.5) into

$$\exists L \in (0, \infty) \text{ and } \delta > 0 \text{ with } S(x) \subseteq S(\bar{x}) + L \mathbf{d}_{\mathscr{X}}(x, \bar{x}) \mathbb{B} \text{ when } \mathbf{d}_{\mathscr{X}}(x, \bar{x}) \leq \delta.$$

From this and Lemma 1.1, we conclude that

$$\mathbf{d}_{\mathrm{exc}}(S(x), S(\bar{x})) \leq L \mathbf{d}_{\mathscr{X}}(x, \bar{x}) \text{ when } x \in \mathrm{dom}(S) \text{ and } \mathbf{d}_{\mathscr{X}}(x, \bar{x}) \leq \delta.$$

($\Leftarrow$) We suppose that S is not calm at $\bar{x}$, so there are sequences of scalars $L^k \to \infty$ and vectors $x^k \to \bar{x}$ as well as $y^k \in S(x^k)$ such that

$$\inf_{y \in S(\bar{x})} \|y^k - y\| > L^k \|x^k - \bar{x}\|. \tag{1.6}$$

From our assumption that S is $\frac{\mathbf{d}_{\mathrm{exc}}}{\mathbf{d}_{\mathscr{X}}}$-calm at $\bar{x}$, we know there exists some $L \in (0, \infty)$ such that for large enough k we have

$$\mathbf{d}_{\mathrm{exc}}\left(S(x^k), S(\bar{x})\right) \leq L \mathbf{d}_{\mathscr{X}}(x^k, \bar{x}) = L \|x^k - \bar{x}\|,$$

where we have applied $x^k \to \bar{x}$ and the fact that $\mathbf{d}_{\mathscr{X}}(x^k, \bar{x}) = \|x^k - \bar{x}\|$ for single vectors. Combining this with (1.6) gives

$$\mathbf{d}_{\mathrm{exc}}\left(S(x^k), S(\bar{x})\right) < \inf_{y \in S(\bar{x})} \|y^k - y\|$$

for large enough k, since the $L^k \to \infty$ will eventually exceed the fixed $L \in (0, \infty)$. This contradicts the definition of $\mathbf{d}_{\mathrm{exc}}\left(S(x^k), S(\bar{x})\right)$ as a supremum over $S(x^k)$ since $y^k \in S(x^k)$. $\quad\square$

As a corollary to Proposition 1.2, we can characterize the localized calmness property when a set-valued mapping S is *calm at $\bar{x}$ for $\bar{y} \in S(\bar{x})$*: there exist $L \in (0, \infty)$, $\delta > 0$, and a neighborhood W of $\bar{y}$ such that

$$S(x) \cap W \subseteq S(\bar{x}) + L \|x - \bar{x}\| \mathbb{B} \quad \text{when } \|x - \bar{x}\| \leq \delta.$$

(For example, see [21, 9(31)] for this definition in the finite-dimensional case. This property was introduced as "upper Lipschitz" continuity in [20].)

Corollary 1.1 *S is calm at $\bar{x}$ for $\bar{y} \in S(\bar{x})$ if and only if there exists a neighborhood W of $\bar{y} \in S(\bar{x})$ such that the W-restricted set-valued mapping:*

$$S_W(x) := \begin{cases} S(x) \cap W & x \neq \bar{x} \\ S(\bar{x}) & x = \bar{x}. \end{cases}$$

is $\frac{\mathbf{d}_{exc}}{\mathbf{d}_{\mathcal{X}}}$-calm at $\bar{x}$ for every pre-distance function $\mathbf{d}_{\mathcal{X}}$.

Proof This follows from Proposition 1.2 applied to S_W.   □

1.3 Derivative Characterization of Calmness

The *upper $\frac{\mathbf{d}_{\mathcal{Y}}}{\mathbf{d}_{\mathcal{X}}}$-derivative of S* takes multisets $\bar{X} \in \mathrm{dom}(S)$ to (non-negative) extended real values as follows:

$$D^+_{\frac{\mathbf{d}_{\mathcal{Y}}}{\mathbf{d}_{\mathcal{X}}}} S(\bar{X}) := \sup_{X^k \searrow^S \bar{X}} \ \limsup_{k \to \infty} \frac{\mathbf{d}_{\mathcal{Y}}\left(S(X^k), S(\bar{X})\right)}{\mathbf{d}_{\mathcal{X}}\left(X^k, \bar{X}\right)},$$

where the notation $X^k \searrow^S \bar{X}$ means that $X^k \in \mathrm{dom}(S)$ and that the pre-distances $\mathbf{d}_{\mathcal{X}}\left(X^k, \bar{X}\right)$ converge to zero from above. If no such sequence of X^k exists, the value of the upper $\frac{\mathbf{d}_{\mathcal{Y}}}{\mathbf{d}_{\mathcal{X}}}$-derivative of S at $\bar{X}$ is defined to be zero. If the two pre-distance functions $\mathbf{d}_{\mathcal{X}}$ and $\mathbf{d}_{\mathcal{Y}}$ are the same type $\mathbf{d}$, we just write $D^+_{\mathbf{d}} S(\bar{X})$.

Proposition 1.3 *S is $\frac{\mathbf{d}_{\mathcal{Y}}}{\mathbf{d}_{\mathcal{X}}}$-calm at $\bar{X}$ if and only if*

(i) $\mathbf{d}_{\mathcal{X}}(X, \bar{X}) = 0$ for $X \in \mathrm{dom}(S) \implies \mathbf{d}_{\mathcal{Y}}\left(S(X), S(\bar{X})\right) = 0$, and
(ii) $D^+_{\frac{\mathbf{d}_{\mathcal{Y}}}{\mathbf{d}_{\mathcal{X}}}} S(\bar{X}) < \infty$.

In this situation, we also know that

$$D^+_{\frac{\mathbf{d}_{\mathcal{Y}}}{\mathbf{d}_{\mathcal{X}}}} S(\bar{X}) = \inf_{\delta > 0}\{\text{calmness moduli } L \text{ with calmness radius } \delta\}. \qquad (1.7)$$

Proof

$(\Rightarrow)$ Condition (i) follows immediately from the definition of $\frac{\mathbf{d}_{\mathcal{Y}}}{\mathbf{d}_{\mathcal{X}}}$-calmness. The condition (ii) follows trivially if there are no sequences with $X^k \in \mathrm{dom}(S)$, $\mathbf{d}_{\mathcal{X}}\left(X^k, \bar{X}\right) > 0$, and $\mathbf{d}_{\mathcal{X}}\left(X^k, \bar{X}\right) \to 0$, since then the upper $\frac{\mathbf{d}_{\mathcal{Y}}}{\mathbf{d}_{\mathcal{X}}}$-derivative is defined to be zero in that case. Otherwise, condition (ii) follows immediately from the definition of $\frac{\mathbf{d}_{\mathcal{Y}}}{\mathbf{d}_{\mathcal{X}}}$-calmness since $\mathbf{d}_{\mathcal{X}}\left(X^k, \bar{X}\right) > 0$ means that we can rewrite the calmness inequality in (1.3) as the ratio (1.4).

($\Leftarrow$) Because of condition (i), we need only to verify $\frac{\mathbf{d}_{\mathscr{Y}}}{\mathbf{d}_{\mathscr{X}}}$-calmness for $X \in$ dom(S) with $\mathbf{d}_{\mathscr{X}}\left(X, \bar{X}\right) > 0$. We proceed by contradiction, supposing there are sequences of scalars $L^k \to \infty$ and multisets $X^k \in$ dom(S), with $\mathbf{d}_{\mathscr{X}}\left(X^k, \bar{X}\right) > 0$ and $\mathbf{d}_{\mathscr{X}}\left(X^k, \bar{X}\right) \to 0$, such that

$$\frac{\mathbf{d}_{\mathscr{Y}}\left(S(X^k), S(\bar{X})\right)}{\mathbf{d}_{\mathscr{X}}(X^k, \bar{X})} > L^k.$$

From this, we conclude $D^{+}_{\frac{\mathbf{d}_{\mathscr{Y}}}{\mathbf{d}_{\mathscr{X}}}} S(\bar{X}) = \infty$ which contradicts condition (ii).

To show the equation (1.7), we first notice that the ratio (1.4) implies that the bound $D^{+}_{\frac{\mathbf{d}_{\mathscr{Y}}}{\mathbf{d}_{\mathscr{X}}}} S(\bar{X}) \leq L$ holds for any calmness modulus L. It follows immediately that $D^{+}_{\frac{\mathbf{d}_{\mathscr{Y}}}{\mathbf{d}_{\mathscr{X}}}} S(\bar{X})$ does not exceed the infimum in (1.7). On the other hand, if there exists an $\tilde{L}$ satisfying

$$D^{+}_{\frac{\mathbf{d}_{\mathscr{Y}}}{\mathbf{d}_{\mathscr{X}}}} S(\bar{X}) < \tilde{L} < \inf_{\delta \to 0} \{\text{calmness moduli } L \text{ with calmness radius } \delta\}, \qquad (1.8)$$

then it follows from the second bound in (1.8) that there must be multisets $X^k \in$ dom(S), with $\mathbf{d}_{\mathscr{X}}\left(X^k, \bar{X}\right) > 0$ and $\mathbf{d}_{\mathscr{X}}\left(X^k, \bar{X}\right) \to 0$, such that

$$\frac{\mathbf{d}_{\mathscr{Y}}\left(S(X^k), S(\bar{X})\right)}{\mathbf{d}_{\mathscr{X}}(X^k, \bar{X})} > \tilde{L};$$

in order that the calmness condition (1.4) be violated for $\tilde{L}$ with every calmness radius $\delta > 0$ (note that we have used (i) to rule out $\mathbf{d}_{\mathscr{X}}(X^k, \bar{X}) = 0$). In this event, we know that $D^{+}_{\frac{\mathbf{d}_{\mathscr{Y}}}{\mathbf{d}_{\mathscr{X}}}} S(\bar{X}) \geq \tilde{L}$, which contradicts the first bound in (1.8). We conclude that the equation (1.7) holds as claimed. $\square$

Remark 1.2 From (1.1), it follows that $\mathbf{d}_{\text{inf}}$ and $\mathbf{d}_{\text{exc}}$ each satisfy condition (i) as $\mathbf{d}_{\mathscr{Y}}$ (with any pre-distance function $\mathbf{d}_{\mathscr{X}}$) in the special case when S is a set-valued mapping (since then $X = \{x\}$ and $\bar{X} = \{\bar{x}\}$, so $\mathbf{d}_{\mathscr{X}}(X, \bar{X}) = 0$ for $X \in$ dom(S) if and only if $x = \bar{x}$), and that $\mathbf{d}_{\text{sup}}$ satisfies condition (i) as $\mathbf{d}_{\mathscr{Y}}$ (with any pre-distance function $\mathbf{d}_{\mathscr{X}}$) in the special case when S is a single-valued mapping.

Remark 1.3 We reference the equation (1.7) by saying that $D^{+}_{\frac{\mathbf{d}_{\mathscr{Y}}}{\mathbf{d}_{\mathscr{X}}}} S(\bar{X})$ is the *limiting calmness modulus for S at* $\bar{X}$. Note that the limiting calmness modulus for S at $\bar{X}$ is not necessarily an actual calmness modulus: for instance, the (single-valued) mapping $S : \mathbb{R} \to \mathbb{R}$ defined by $S(x) = \{x^2\}$ is $\mathbf{d}$-calm at 0 for any calmness radius $\delta > 0$, but $L = \delta$ is the smallest possible corresponding calmness modulus, though the limiting calmness modulus for S at 0 is $D^{+}_{\mathbf{d}} S(0) = 0$.

One of our results in the next chapter will require that a multiset mapping S is $\frac{d\mathscr{Y}}{d\mathscr{X}}$-*contractive at* $\bar{X}$, which means that S is $\frac{d\mathscr{Y}}{d\mathscr{X}}$-calm at $\bar{X}$ with calmness modulus $L < 1$. If the same property holds but with $L \leq 1$ (instead of the strict inequality), we say that S is $\frac{d\mathscr{Y}}{d\mathscr{X}}$-*non-expansive at* $\bar{X}$. The following corollary is an immediate consequence of Proposition 1.3.

Corollary 1.2 S is $\frac{d\mathscr{Y}}{d\mathscr{X}}$-*contractive at* $\bar{X}$ *if and only if:*

(i) $\mathbf{d}_{\mathscr{X}}(X, \bar{X}) = 0$ *for* $X \in \mathrm{dom}(S) \implies \mathbf{d}_{\mathscr{Y}}\left(S(X), S(\bar{X})\right) = 0$, *and*

(ii) $D^{+}_{\frac{d\mathscr{Y}}{d\mathscr{X}}} S(\bar{X}) < 1$.

Chapter 2
Iteration Mappings

A minimization method $\mathscr{M}$ applied to an objective function $f : \mathbb{R}^n \to \mathbb{R} \cup \{\infty\}$ defines a special multiset mapping $I_{\mathscr{M},f} : \mathbb{R}^n \rightrightarrows \mathbb{R}^n$ called an *iteration mapping*. Iteration mappings operate on *iterate-multisets* $X \subseteq \mathbb{R}^n$ with fixed cardinality $|X| = m \geq 1$ (where m is determined by the particular minimization method $\mathscr{M}$), and the image under $I_{\mathscr{M},f}$ of an iterate-multiset X is either another iterate-multiset or the empty multiset.

From an initial iterate-multiset $X^0 \in \text{dom}\left(I_{\mathscr{M},f}\right)$, we get a new iterate-multiset $X^1 = I_{\mathscr{M},f}\left(X^0\right)$ via the iteration mapping. Likewise, as long as $X^1 \in \text{dom}\left(I_{\mathscr{M},f}\right)$, we get a new iterate-multiset $X^2 = I_{\mathscr{M},f}\left(X^1\right)$. As long as all of the new iterate-multisets are in $\text{dom}\left(I_{\mathscr{M},f}\right)$, we can continue in this manner to generate a sequence $\{X^k\}$ of iterate-multisets from the iteration mapping via:

$$X^k = I_{\mathscr{M},f}\left(X^{k-1}\right) = I_{\mathscr{M},f}^{(k)}\left(X^0\right),$$

where $I_{\mathscr{M},f}^{(k)}$ is shorthand for k compositions of the iteration mapping:

$$I_{\mathscr{M},f}^{(k)}(X) := \underbrace{I_{\mathscr{M},f} \circ I_{\mathscr{M},f} \circ \ldots \circ I_{\mathscr{M},f}}_{k \text{ times}}(X).$$

Accordingly, we define the collection $\mathscr{V}_{\mathscr{M},f}$ of all *viable* iterate-multisets:

$$\mathscr{V}_{\mathscr{M},f} := \{X \in \text{dom}\left(I_{\mathscr{M},f}\right) : I_{\mathscr{M},f}^{(k)}(X) \in \text{dom}\left(I_{\mathscr{M},f}\right) \text{ for all } k \geq 1\}.$$

When $m = 1$, we simply refer to the iterate-multisets as *iterates* and denote them in lower case x^k.

To illustrate what iteration mappings can look like, we construct them for coordinate-search, steepest-descent (with or without a line-search), and Nelder–Mead. To describe these minimization methods, we will use the terminology of

© The Author(s), under exclusive licence to Springer Nature Switzerland AG 2018
A. B. Levy, *Attraction in Numerical Minimization*, SpringerBriefs in Optimization,
https://doi.org/10.1007/978-3-030-04049-9_2

"better" and "worse" as shorthand for "has lower f-value" and "has higher f-value" (respectively) when comparing two points. Note that the iteration mappings below represent each of the methods in a way that is well suited to our analysis, though the direct iteration of these mappings would not necessarily be the most efficient implementation of the methods.

2.1   Coordinate-Search

This is a pattern-search method that begins by testing trial points $\{x^k \pm e_i \ : \ i = 1, \ldots, n\}$ in the coordinate directions around the current iterate x^k (e_i denotes the n-vector of all zeroes except for a 1 in the ith entry). When one of these trial points is better than the current iterate x^k, it becomes the new iterate (Figure 2.1). If none of these trial points is better than x^k, the method backtracks halfway along the same pattern and tests the new set of trial points. The method continues backtracking in this way until a better trial point is found.

2.1.1   Iteration Mapping $I_{CS,f}$

The coordinate-search method generates iterates via the iteration mapping:

$$I_{CS,f}(x) := \mathrm{First}\left(\underset{y \in T(x)}{\mathrm{argmin}} \, f(y)\right)$$

Fig. 2.1 Coordinate-Search for $n = 2$. The trial point to the north is better

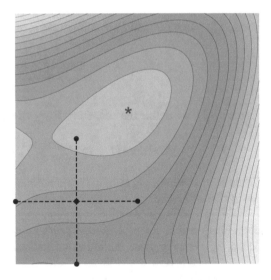

where the ordered trial set $T(x)$ is defined by:

$$T(x) := \{x, \; x + \alpha(x) e_1, \; x - \alpha(x) e_1, \ldots, x + \alpha(x) e_n, \; x - \alpha(x) e_n\},$$

in terms of the step-size function:

$$\alpha(x) := \max \left\{ \operatorname*{argmax}_{\alpha \in \{2^{-j} : j \in \mathbb{N}\}} \delta_{[1,\infty)} \left(\sum_{i=1}^{n} \left(\delta_{(0,\infty]} (f(x) - f(x + \alpha e_i)) + \delta_{(0,\infty]} (f(x) - f(x - \alpha e_i)) \right) \right) \right\}.$$

Remark 2.1 If x minimizes f on $\mathbb{B}(x; 1)$, then $\alpha(x) = 1$ and x is fixed by this iteration mapping: $I_{CS,f}(x) = x$.

Remark 2.2 The collection of viable iterates $\mathcal{V}_{CS,f}$ is the entire space $\mathbb{R}^n$ regardless of the objective function.

2.2 Steepest-Descent

This classic method assumes that f is differentiable and tests a trial point $x^k - \alpha \nabla f(x^k)$ in the direction of steepest descent from the current iterate x^k. This method is usually augmented by a line-search to ensure that a better trial point is eventually found if it exists.

2.2.1 Iteration Mapping $I_{SD_i, f}$

The steepest-descent method generates iterates via the iteration mapping:

$$I_{SD_i, f}(x) = x - \alpha_i(x) \nabla f(x)$$

in terms of the following step-size functions:

- Without line-search: $\alpha_1(x) = 1$,
- With exact line-search:

$$\alpha_2(x) = \min \left(\operatorname*{argmin}_{\alpha \geq 0} f(x - \alpha \nabla f(x)) \right),$$

- With no-longer-downhill line-search: $\alpha_3(x) = 1$ when $\nabla f(x) = 0$, and otherwise:

$$\alpha_3(x) = \min \left(\operatorname*{argmax}_{\alpha \geq 0} \delta_{(0,\infty)} \left(f(x) - f(x - \alpha \nabla f(x)) \right) \right),$$

- With backtracking line-search: $\alpha_4(x) = 1$ when $\nabla f(x) = 0$, and otherwise:

$$\alpha_4(x) = \max \left(\operatorname*{argmax}_{\alpha \in \{2^{-j} : j \in \mathbb{N}\}} \delta_{(0,\infty)} \left(f(x) - f(x - \alpha \nabla f(x)) \right) \right).$$

Remark 2.3 If we do not require strict descent, we can use the indicator of the non-negative numbers $[0, \infty)$ instead in either of the final two line-searches. On the other hand, we can require "sufficient" descent by instead including an Armijo–Goldstein term like $-\alpha \beta \|\nabla f(x)\|^2$ inside the indicator to guarantee that the reduction in f-value is at least as good as $\beta \in (0, 1)$ times the reduction predicted by the first-order Taylor approximation.

Remark 2.4 In all cases, $I_{\mathrm{SD}_i, f}(x) = x$ if $\nabla f(x) = 0$. Since $\nabla f(x) = 0$ is a necessary condition for x to be a local minimizer of f, we have that any local minimizer x of f is fixed by this iteration mapping: $I_{\mathrm{SD}_i, f}(x) = x$.

Remark 2.5 If (i) $\nabla f(\bar{x}) = 0$, (ii) ∇f is calm at $\bar{x}$ (with radius $\delta > 0$, and modulus L), and

$$\text{(iii)} \qquad \sup_{\{x : \|x - \bar{x}\| \leq \delta\}} |\alpha_i(x)| \leq \bar{\alpha},$$

then for any $x \in \{x : \|x - \bar{x}\| \leq \delta\}$ we have

$$\|I_{\mathrm{SD}_i, f}(x) - I_{\mathrm{SD}_i, f}(\bar{x})\| = \|x - \alpha_i(x) \nabla f(x) - \bar{x}\|$$
$$= \|x - \alpha_i(x) \nabla f(x) - \bar{x} + \alpha_i(x) \nabla f(\bar{x})\|$$
$$\leq (1 + |\alpha_i(x)| L) \|x - \bar{x}\| \leq (1 + \bar{\alpha} L)\|x - \bar{x}\|;$$

in which case the steepest-descent iteration mapping $I_{\mathrm{SD}_i, f}$ is calm at $\bar{x}$ with radius δ and modulus $1 + \bar{\alpha} L$. Notice for example that the backtracking line-search α_4 satisfies condition (iii) with $\bar{\alpha} = 1$.

Remark 2.6 Viability of the initial iterate does not ensure convergence (or even boundedness) of the sequence of iterates, as we will illustrate with the following two examples:

- $f(x_1, x_2) = x_1^2 + x_2^2$ with SD$_1$: the iteration mapping satisfies $I_{\mathrm{SD}_1, x_1^2 + x_2^2}(x) = -x$ and the collection of viable iterates $\mathcal{V}_{\mathrm{SD}_1, x_1^2 + x_2^2}$ is the entire space $\mathbb{R}^2$. For any non-zero initial iterate x^0, the sequence of iterates x^k alternates between $-x^0$ and x^0.
- $f(x_1, x_2) = \frac{1}{x_1^2 + x_2^2}$ with SD$_1$ (or SD$_4$): the iteration mapping satisfies

$$I_{\mathrm{SD}_1, \frac{1}{x_1^2 + x_2^2}}(x) = \left(2 \, (f(x))^2 + 1\right) x$$

and the collection of viable iterates $\mathcal{V}_{\mathrm{SD}_1, \frac{1}{x_1^2 + x_2^2}}$ is $\mathrm{dom}(f) = \mathbb{R}^2 \setminus (0,0)$. The iterates x^k grow without bound $\|x^k\| \to \infty$ from any viable initial iterate x^0.

Fig. 2.2 Outer contract

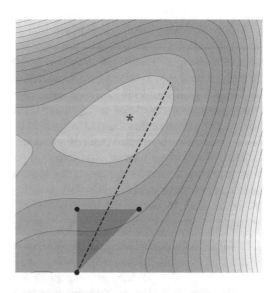

Fig. 2.3 Shrink toward best vertex

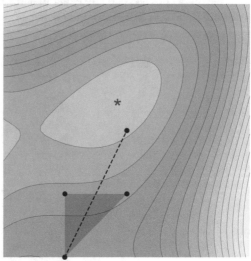

2.3   Nelder–Mead

This method is categorized as a "simplex method" because it uses an *n-simplex*, which is any multiset $X \subseteq \mathbb{R}^n$ with cardinality $|X| = n+1$. The individual elements x in a simplex X are called "vertices" in this context; so, for example, the three vertices of any triangle in $\mathbb{R}^2$ form a 2-simplex. A non-degenerate n-simplex is one where the edges connecting any vertex of the n-simplex to each of the other vertices form a basis for $\mathbb{R}^n$ (e.g., when the vertices of a triangle are not on the same line).

The Nelder–Mead method generates trial points along the ray from the worst vertex through the "centroid" of the opposite face (e.g., the midpoint of the opposite edge of a triangle). Figure 2.4 shows an example of a 2-simplex (the vertices in purple), as well as part of the ray (the dotted purple line segment) from the worst vertex through the centroid of the opposite edge: The idea is that by moving along this ray away from the worst point, we are likely to be moving toward better points (simulating a descent direction). The method is named after its authors Nelder and Mead [19]; however, we will describe a version outlined in, e.g., [15].

The Nelder–Mead method tests four trial points in turn along the ray from the worst vertex and shrinks the n-simplex toward the best vertex if those four trial points do not prove sufficient. The following is an illustrated outline of the method in two dimensions, where the first trial point is the reflection of the worst vertex along the ray through the opposite face as in Figure 2.5.

1. If the reflect point is better than the second-best but no better than the best, swap it for the worst vertex to create a new n-simplex. The rationale is that the reflect point is a good enough improvement to use.
2. If the reflect point is actually better than the best, consider the expand trial point as in Figure 2.6. Then, swap the better of the expand or the reflect points with

Fig. 2.4 A 2-simplex and ray from worst vertex

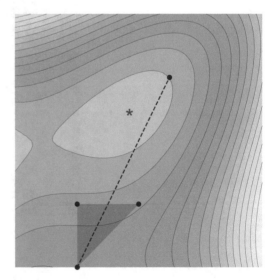

the worst vertex to create a new n-simplex. (This is precisely what would happen in the case illustrated in Figure 2.6, where the reflect point would be chosen.) The rationale is that the reflect point is such a good improvement that a more aggressive step in its direction might be even better.

3. If the reflect point is no better than the second-best vertex, but better than the worst vertex, test the outer contract trial point as in Figure 2.2. If the outer contract point is at least as good as the reflect point, swap it for the worst vertex. Otherwise, shrink the n-simplex toward the best vertex as in Figure 2.3. The rationale is that the reflect point is not a good enough improvement to keep, but

Fig. 2.5 Reflection of worst vertex across opposite face

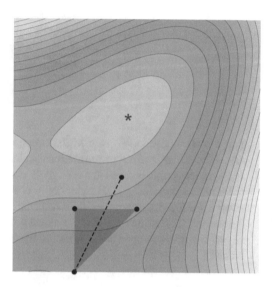

Fig. 2.6 Expand

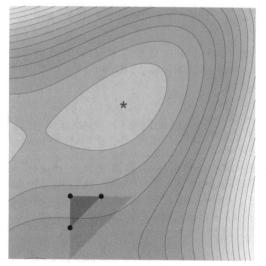

Fig. 2.7 Inner contract

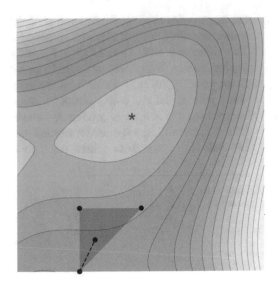

that a more conservative step in its direction might be. If not, the shrink step resets the n-simplex at a less ambitious scale.

4. If the reflect point is no better than the worst point, test the inner contract trial point as in Figure 2.7. If the inner contract point is better than the worst vertex, swap it for the worst vertex. Otherwise, shrink the n-simplex toward the current best vertex. The rationale is that the reflect point is definitely not an improvement, so we consider an even more conservative step in its direction than the outer contract.

2.3.1   Iteration Mapping $I_{\mathrm{NM},f}$

The iterate-multisets $X \subseteq \mathbb{R}^n$ generated by the Nelder–Mead method are (ordered) n-simplices. To incorporate ordering into the Nelder–Mead iteration mapping, we define

$$x_i(X) = \mathrm{First}\left(\operatorname*{argmin}_{x \in X \backslash \{x_j(X): j < i\}} f(x) \right) \quad \text{for } i = 0, 1, \ldots, n,$$

where multiple elements in each argmin set are ordered canonically as n-vectors (by smallest first component, then by smallest second component if same first component, etc.). Note that the "First" mapping in the definition of x_i ensures that $x_i(X)$ is single valued, which is necessary to ensure a centroid:

$$C(X) = \frac{1}{n} \sum_{i=1}^{n} x_i(X)$$

that is well-defined (e.g., $x_n(X)$ would be empty if more than one element of X was extracted into any earlier set $x_i(X)$). We also use trial point mappings defined by:

$$R(X) = C(X) + \rho\ (C(X) - x_n(X))$$

$$E(X) = C(X) + \chi\ (R(X) - C(X))$$

$$O(X) = C(X) + \gamma\ (R(X) - C(X))$$

$$I(X) = C(X) + \gamma\ (x_n(X) - C(X))$$

$$S(X) = \left\{ x_0(X), x_0(X) + \sigma\ (x_1(X) - x_0(X)), \ldots, x_0(X) + \sigma\ (x_n(X) - x_0(X)) \right\}$$

The standard choice of parameters is $\rho = 1$, $\chi = 2$, and $\gamma = \sigma = \frac{1}{2}$, and the notation translates to R (reflect), E (expand), O (outer contract), I (inner contract), and S (shrink).

The Nelder–Mead iteration mapping is then given by:

$$I_{\mathrm{NM},f}(X) := T_E(X) + T_R(X) + T_O(X) + T_I(X) + T_S(X),$$

for the trial multiset mappings defined as follows:

$$T_E(X) := \delta_{\left[-\infty, f_0\right)}(f_R)\ \mathrm{pos}\,(f_R - f_E)\ \{x_0(X), x_1(X), \ldots, x_{n-1}(X), E(X)\}$$

$$T_R(X) := \delta_{\left[-\infty, f_0\right)}(f_R)\ \mathrm{posc}\,(f_R - f_E)\ \{x_0(X), x_1(X), \ldots, x_{n-1}(X), R(X)\}$$

$$T_O(X) := \delta_{\left[f_{n-1}, f_n\right)}(f_R)\ \mathrm{posc}\,(f_O - f_R)\ \{x_0(X), x_1(X), \ldots, x_{n-1}(X), O(X)\}$$

$$T_I(X) := \delta_{\left[f_n, \infty\right]}(f_R)\ \delta_{\left[-\infty, f_n\right)}(f_I)\ \{x_0(X), x_1(X), \ldots, x_{n-1}(X), I(X)\}$$

$$T_S(X) := \Big(\delta_{\left[f_0, f_{n-1}\right)}(f_R) + \delta_{\left[f_{n-1}, f_n\right)}(f_R)\ \mathrm{pos}\,(f_O - f_R) + \delta_{\left[f_n, \infty\right]}(f_R)$$

$$\delta_{\left[f_n, \infty\right]}(f_I) \Big)\ S(X)$$

in terms of:

$$\mathrm{pos}(a) = \delta_{(0,\infty]}(a) = \begin{cases} 1 \text{ if } a \in (0, \infty] \\ 0 \text{ if } a \notin (0, \infty] \end{cases}$$

$$\mathrm{posc}(a) = 1 - \mathrm{pos}(a) = \begin{cases} 0 \text{ if } a \in (0, \infty] \\ 1 \text{ if } a \notin (0, \infty] \end{cases}$$

$$f_i = f(x_i(X))$$

$$f_R = f(R(X))$$

$$f_E = f(E(X))$$

$$f_O = f(O(X))$$
$$f_I = f(I(X))$$

Note that there is an alternate representation of the indicator functions in the trial multiset mappings in terms of pos and posc composed with f as follows:

$$\delta_{[-\infty, f_0)}(f_R) = \begin{cases} 1 \text{ if } f_R < f_0 \\ 0 \text{ otherwise} \end{cases}$$
$$= \text{pos}\left(f(x_0(X)) - f(R(X))\right)$$

$$\delta_{[f_{n-1}, f_n)}(f_R) = \begin{cases} 1 \text{ if } f_{n-1} \le f_R < f_n \\ 0 \text{ otherwise} \end{cases}$$
$$= \text{posc}\left(f(x_{n-1}(X)) - f(R(X))\right) \text{pos}\left(f(x_n(X)) - f(R(X))\right)$$

$$\delta_{[f_n, \infty]}(f_R) = \begin{cases} 1 \text{ if } f_n \le f_R \\ 0 \text{ otherwise} \end{cases}$$
$$= \text{posc}\left(f(x_n(X)) - f(R(X))\right)$$

$$\delta_{[-\infty, f_n)}(f_I) = \begin{cases} 1 \text{ if } f_I < f_n \\ 0 \text{ otherwise} \end{cases}$$
$$= \text{pos}\left(f(x_n(X)) - f(I(X))\right)$$

$$\delta_{[f_1, f_n)}(f_R) = \begin{cases} 1 \text{ if } f_1 \le f_R < f_n \\ 0 \text{ otherwise} \end{cases}$$
$$= \text{posc}\left(f(x_1(X)) - f(R(X))\right) \text{pos}\left(f(x_n(X)) - f(R(X))\right)$$

Remark 2.7 Any multiset $X^\infty \subseteq \mathbb{R}^n$ consisting of $n+1$ copies of the same element $x^\infty \in \mathbb{R}^n$ is fixed by this iteration mapping: $I_{NM, f}(X^\infty) = X^\infty$.

Remark 2.8 The collection of viable iterate-multisets $\mathcal{V}_{NM, f}$ for any objective function f is the collection of vertices of non-degenerate n-simplices.

2.4   Practical Considerations

In practice, a minimization method $\mathcal{M}$ applied to an objective function f always terminates after a finite number of iterations for one of the two reasons:

1. A stopping criterion is satisfied by the current iterate-multiset, or
2. A new iterate-multiset cannot be determined from the current iterate-multiset.

In order to study asymptotic behavior, we need to adopt conventions for iteration mappings under these circumstances. In particular, we handle the first circumstance with the convention that any iterate-multiset X satisfying the stopping criterion associated with the method $\mathcal{M}$ applied to f is fixed by the iteration mapping $I_{\mathcal{M},f}(X) = X$. We handle the second circumstance with the convention that any iteration mapping applied to the empty set returns the empty set.

Finally, if any iterate-multiset:

$$X^\infty := \underbrace{\{x^\infty, x^\infty, \ldots, x^\infty\}}_{m \text{ copies}}$$

consisting of m copies of an attractor (defined in Chapter 4) for the iteration mapping $I_{\mathcal{M},f}$ is not already in the domain of the iteration mapping, we augment the iteration mapping by defining it to fix X^∞: $I_{\mathcal{M},f}(X^\infty) = X^\infty$.

2.5 Convergence Theorems for Iteration Mappings

The next result is an analogue to the classical Banach fixed-point theorem [5, Theorem 6], and assumes an iterate-multiset $\bar{X}$ that is fixed by the iteration mapping: $I_{\mathcal{M},f}(\bar{X}) = \bar{X}$.

Theorem 2.1 *For any pre-distance function $\mathbf{d}$, if the iterate-multiset $\bar{X} \subseteq \mathbb{R}^n$ is fixed by $I_{\mathcal{M},f}$ and $I_{\mathcal{M},f}$ is $\mathbf{d}$-contractive at $\bar{X}$ with radius $\delta > 0$, then $\mathbf{d}(X^k, \bar{X}) \to 0$ as long as the initial iterate-multiset $X^0 \in \mathcal{V}_{\mathcal{M},f}$ satisfies $\mathbf{d}(X^0, \bar{X}) \leq \delta$.*

Proof Since the initial iterate-multiset X^0 satisfies $\mathbf{d}(X^0, \bar{X}) \leq \delta$, we know from (1.3) (with $X = X^0$ and $S = I_{\mathcal{M},f}$) and $I_{\mathcal{M},f}(\bar{X}) = \bar{X}$ that

$$0 \leq \mathbf{d}(X^1, \bar{X}) \leq L\,\mathbf{d}(X^0, \bar{X}) \leq L\,\delta$$

(where the first inequality comes from the fact that pre-distance functions are non-negative on non-empty multisets). Since $L < 1$, this means in particular that $\mathbf{d}(X^1, \bar{X}) \leq \delta$. We can likewise apply condition (1.3) recursively to get

$$0 \leq \mathbf{d}(X^k, \bar{X}) \leq L^k \delta \quad \text{for all } k \geq 0,$$

from which it follows that $\mathbf{d}(X^k, \bar{X}) \to 0$. □

Remark 2.9 We can use Corollary 1.2 to establish that M is $\mathbf{d}$-contractive at $\bar{X}$ via the conditions:

(i) $\mathbf{d}(X, \bar{X}) = 0$ for $X \in \mathrm{dom}(I_{\mathcal{M},f}) \implies \mathbf{d}(I_{\mathcal{M},f}(X), \bar{X}) = 0$, and
(ii) $D_{\mathbf{d}}^+ I_{\mathcal{M},f}(\bar{X}) < 1$.

Remark 2.10 The original Banach fixed-point theorem [5, Theorem 6] assumes a stronger form of contraction via Lipschitz continuity.

The requirement in Theorem 2.1 of **d**-contractiveness can be eased to **d**-non-expansiveness under the additional condition that

$$\liminf_{k \to \infty} \frac{\mathbf{d}\left(I_{\mathscr{M},f}(X^k), \bar{X}\right)}{\mathbf{d}\left(X^k, \bar{X}\right)} < 1. \tag{2.1}$$

Theorem 2.2 *For any pre-distance function* **d**, *if the iterate-multiset* $\bar{X} \subseteq \mathbb{R}^n$ *is fixed by* $I_{\mathscr{M},f}$ *and* $I_{\mathscr{M},f}$ *is* **d**-*non-expansive at* $\bar{X}$ *with radius* $\delta > 0$, *then* $\mathbf{d}(X^k, \bar{X}) \to 0$ *as long as the condition (2.1) holds and the initial iterate-multiset* $X^0 \in \mathscr{V}_{\mathscr{M},f}$ *satisfies* $\mathbf{d}(X^0, \bar{X}) \leq \delta$.

Proof Using the same argument from the proof of Theorem 2.1, we get the inequality:

$$0 \leq \mathbf{d}(X^{k+1}, \bar{X}) \leq \mathbf{d}(X^k, \bar{X})$$

for all $k \geq 0$. This means that the sequence of non-negative scalars defined by $\eta^k := \mathbf{d}(X^k, \bar{X})$ is non-increasing. From condition (2.1), we get that there is a scalar $\tilde{L} < 1$ and (at least) a subsequence for which $\eta^{k+1} \leq \tilde{L}\, \eta^k$. Since $\tilde{L} < 1$ is fixed, we conclude that the full sequence of $\eta^k = \mathbf{d}(X^k, \bar{X})$ converges to zero as claimed. □

Corollary 2.1 *For* $x^\infty \in \mathbb{R}^n$, *if the iterate-multiset* $X^\infty := \{x^\infty, \ldots, x^\infty\}$ *is fixed by* $I_{\mathscr{M},f}$ *and* $I_{\mathscr{M},f}$ *is* $\mathbf{d}_{\mathrm{exc}}$-*non-expansive at* X^∞ *with radius* $\delta > 0$, *then each* $x_i^k \in X^k$ *for* $i = 1, \ldots, m$ *converges to* x^∞ *as long as the condition (2.1) holds with* $\mathbf{d} = \mathbf{d}_{\mathrm{exc}}$ *and the initial iterate-multiset* $X^0 \in \mathscr{V}_{\mathscr{M},f}$ *satisfies* $\mathbf{d}_{\mathrm{exc}}(X^0, X^\infty) \leq \delta$.

Proof This follows immediately from Theorem 2.2 via the definition of the excess pre-distance function $\mathbf{d}_{\mathrm{exc}}$ and the iterate-multiset $X^\infty := \{x^\infty, \ldots, x^\infty\}$ in this case.   □

Remark 2.11 This result is useful when the pre-distance of the iterate-multisets to X^∞ does not necessarily change from one step to the next. Nelder–Mead is an example of a method for which this may happen.

Remark 2.12 Corollary 2.1 holds with $\mathbf{d}_{\mathrm{sup}}$ in place of $\mathbf{d}_{\mathrm{exc}}$ since they are equivalent when applied with the iterate-multiset $X^\infty := \{x^\infty, \ldots, x^\infty\}$ in their second argument.

Chapter 3
Equilibria in Dynamical Systems

Dynamical systems describe how a state develops over time, and differential equations are an example of a dynamical system where time is measured continuously. The reader can see any introductory text on dynamical systems or differential equations (e.g., [7]) for many more details, but we'll illustrate some of the concepts that have analogues in our work by considering the following (coupled) differential equations:

$$x_1' = -1 - x_1^2 + x_2$$
$$x_2' = 1 + x_1 - x_2^2. \tag{3.1}$$

The states associated with this dynamical system are pairs of functions $(x_1(t), x_2(t))$, parameterized by time $t \in [0, \infty)$, that simultaneously satisfy both differential equations. More general systems of differential equations can be expressed as $x' = F(x)$, where $x : [0, \infty) \to \mathscr{R}^n$ represents a vector-valued state function of the parameter $t \in [0, \infty)$, and $F : \mathscr{R}^n \to \mathscr{R}^n$ is a mapping defining the system.

The *phase space* for the dynamical system (3.1) is the $x_1 x_2$-plane pictured in Figure 3.1, where the directed curves (called *trajectories*) show the evolution of the state pair $(x_1(t), x_2(t))$. Different trajectories in the phase plane correspond to different initial states $(x_1(0), x_2(0))$, and numerical methods (e.g., Euler's method) for approximating the evolution of the state from $(x_1(0), x_2(0))$ generate a sequence of pairs (x_1^k, x_2^k) approximately on the trajectory through $(x_1(0), x_2(0))$. More generally, numerical methods for approximating the evolution of states associated with $x' = F(x)$ generate sequences of vectors $x^k \in \mathscr{R}^n$, much as iterate-multisets X^k are generated by a minimization method.

The *equilibria* of a system of differential equations $x' = F(x)$ are the vectors x^∞ for which $F(x^\infty) = 0$. In the example (3.1), there are two equilibria, one at $(0, 1)$ and one at $(0.45, 1.2)$, shown as stars in blue and red, respectively, in Figure 3.1.

© The Author(s), under exclusive licence to Springer Nature Switzerland AG 2018
A. B. Levy, *Attraction in Numerical Minimization*, SpringerBriefs in Optimization,
https://doi.org/10.1007/978-3-030-04049-9_3

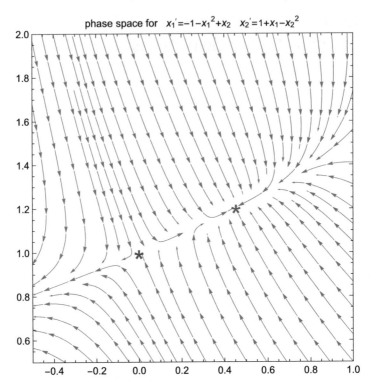

phase space for $x_1' = -1 - x_1{}^2 + x_2$ $x_2' = 1 + x_1 - x_2{}^2$

Fig. 3.1 The two equilibria for the dynamical system (3.1)

Equilibria in dynamical systems motivate our definition in the next chapter of "attractors" associated with the numerical minimization of an objective function. A given dynamical system entirely determines its equilibria; however, as we will see, attractors are determined not only by the objective function (effectively, the "system" in a minimization problem) but also by the numerical minimization method used.

3.1   Basins of Attraction

The *basin of attraction* associated with an equilibrium x^∞ is the set Basin$(x^\infty) \subseteq \mathscr{R}^n$ of initial states whose trajectories are attracted to x^∞. For instance, the basin of attraction for the equilibrium $(0, 1)$ of the dynamical system (3.1) would consist of all the points on the two blue trajectories leading to $(0, 1)$ in Figure 3.2 as well as the equilibrium at $(0, 1)$ itself. The basin of attraction of the equilibrium $(0.45, 1.2)$ would be all the points shaded red in Figure 3.2, to the right of the basin of attraction of the equilibrium at $(0, 1)$.

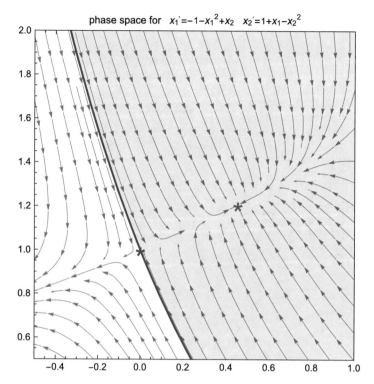

phase space for $x_1' = -1 - x_1^2 + x_2$ $x_2' = 1 + x_1 - x_2^2$

Fig. 3.2 Basins of attraction

In the next chapter, we will define basins of attraction associated with the numerical minimization of an objective function, and we will use simulation to identify and analyze these objects (there is precedent for this type of simulation in [3]). For comparison, we simulated the basins shown in Figure 3.2 by pseudo-randomly choosing 10,000 initial points within the portion of the phase plane shown and color-coded them according to the equilibrium to which the resulting trajectory was attracted. The results of these simulations are shown in Figure 3.3, where we see that the basin for the (blue) equilibrium $(0, 1)$ does not appear. This absence is not surprising since pseudo-random sampling in a two-dimensional space is not expected to locate a one-dimensional object.

3.1.1 Basin Size

The authors of [22] use simulation to classify the size of a basin of attraction associated with an equilibrium according to the model $P(r) = P_0 r^{-\gamma}$ for the asymptotic behavior of the probability that an initial condition within a distance r of the equilibrium is also in its basin of attraction. Their procedure first estimates

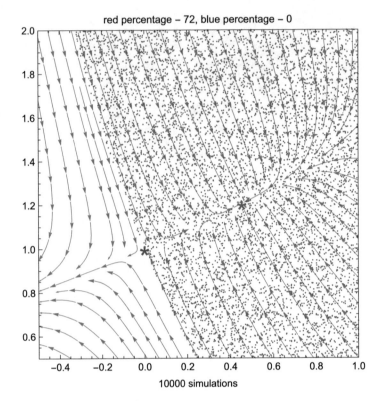

Fig. 3.3 Simulated basins of attraction

$P(1)$ by pseudo-randomly choosing initial points uniformly distributed in the n-hypersphere of radius $r = 1$ about the equilibrium (where n is the dimension of the phase space), and calculates the fraction of them in the basin of attraction. Then, the same process is used to estimate $\Delta P(1)$ by sampling within the n-hypershell between the n-hyperspheres of radii $r = 1$ and $r = 2$ about the equilibrium. This estimate is combined with that of $P(1)$ to approximate $P(2)$ according to:

$$P(2) = 2^{-n} P(1) + \left(1 - 2^{-n}\right) \Delta P(1),$$

where the weightings on each term account for the different volumes of the hyperspheres. This process is repeated to establish estimates of $P(2^d)$ for increasing integer powers d.

We applied the size-estimating procedure of [22] to the two equilibria in (3.1) (using 1000 samples at each step, and d up to 10), and got $P(2^d) = 0$ for the equilibrium at $(0, 1)$ for all d (this is again consistent with the fact that the basin of attraction in this case is one-dimensional). For the other equilibrium at $(0.45, 1.2)$, we got the estimates shown in Table 3.1.

Table 3.1 Basin size data for equilibrium (0.45, 1.2)

d	0	1	2	3	4	5	6	7	8	9	10
$P(2^d)$	0.798	0.620	0.505	0.407	0.350	0.303	0.280	0.253	0.266	0.243	0.239

We fit a line to the last four points on the $\log_2$ plot of the probability data in Table 3.1 to determine the values of $\gamma = 0$ and $P_0 = 0.31$ in the asymptotic model $P(r) = P_0 r^{-\gamma}$.

The main size-classes identified in [22] are as follows (from largest to smallest):

1. $\gamma = 0$ and $P_0 = 1$; the equilibrium attracts trajectories from almost every initial point.
2. $\gamma = 0$ and $P_0 < 1$; the equilibrium attracts trajectories from $\approx P_0$ of all initial points.
3. $\gamma \in (0, n)$; the basin has co-dimension γ.
4. $\gamma = n$; the basin is bounded with effective radius $r_0 := P_0^{\frac{1}{n}}$.

Our results above indicate that the basin of attraction for the equilibrium at (0.45, 1.2) is in the second class, and so should attract approximately 31% of all initial points. If we expand our view of the phase space as in Figure 3.4, we indeed see that the basin appears to fill slightly more than a quarter of the space. An even more accurate value for P_0 may be obtained by using a greater number of samples and by generating the data for larger integer values of d.

In the final chapter, we will adapt the classification of [22] to make our own classification of basin size for basins of attraction associated with the numerical minimization of an objective function, and we will use this classification as one component of our investigation of the practical significance of two well-known counterexamples to good behavior in numerical minimization.

3.1.2 Basin Entropy

The state in a dynamical system can evolve from nearby initial points very differently or very similarly, depending on how those initial points relate to the basins of attraction for the system. For instance, a state initiated on either of the two blue trajectories leading to (0, 1) in Figure 3.2 will always evolve to the equilibrium at (0, 1), while a state initiated just to the right will instead evolve to the equilibrium at (0.45, 1.2) and a state initiated just to the left will evolve down and to the left in the phase space, asymptotic to the line through the two equilibria. On the other hand, any pair of states initiated inside the red basin will evolve to the equilibrium at (0.45, 1.2). In this sense, it is harder to predict the evolution of a state initiated near the two blue trajectories than it is to predict the evolution of a state initiated well inside the red basin.

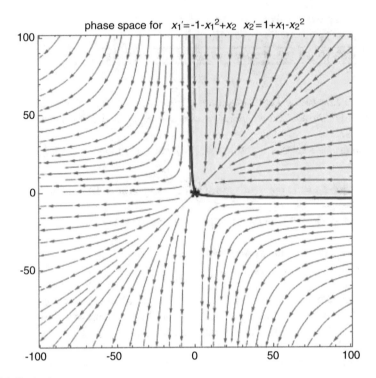

Fig. 3.4 Basin sizes

The authors of [9] develop a notion of "basin entropy" to quantify the cumulative unpredictability of this kind within a region Ω of the phase space. Their first step is to subdivide Ω into a grid of N equal boxes (n-hypercubes with edge-length ϵ). They then compute the probability $p_{i,j}$ of each evolutionary outcome j (out of the J_i possible outcomes in box i) as the fraction of 25 different trajectories initiated pseudo-randomly in box i that evolve to outcome j. We have illustrated the results of this process for the dynamical system (3.1) in the image on the right of Figure 3.5, where a darker shade of red indicates a higher probability that the trajectories in that box evolve to the red equilibrium at (0.45, 1.2). As before, none of the pseudo-random initial points generate trajectories evolving to the blue equilibrium at (0, 1).

The *basin entropy* Sb is defined in [9] to be the average Gibbs entropy over all N boxes:

$$Sb := \frac{1}{N} \sum_{i=1}^{N} \sum_{j=1}^{J_i} p_{i,j} \log\left(\frac{1}{p_{i,j}}\right), \tag{3.2}$$

and a low basin entropy suggests relatively simple basin boundaries (i.e., not fractal boundaries). The image on the left in Figure 3.5 shows darker shades for boxes with higher Gibbs entropy, and we see those aligning with the boundary of the

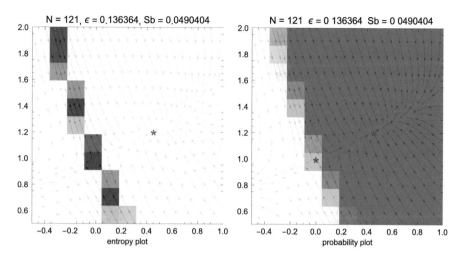

Fig. 3.5 Entropy and probability

basin of attraction for the red equilibrium at (0.45, 1.2) (consistent with our earlier discussion of unpredictability for this example). We also see that the relatively small basin entropy of $Sb = 0.0519462$ is consistent with the fact that this example has a very simple basin boundary.

In the final chapter, we will develop and calculate a similar measurement of basin entropy for basins of attraction associated with the numerical minimization of an objective function, and we will use basin entropy as another component of our investigation of the practical significance of two well-known counterexamples to good behavior in numerical minimization.

3.2 Stability

It is evident from Figure 3.6 that every state initiated within the red circle stays within that circle, and that this is not the case with the blue circle. This illustrates the following stability property attributed to Lyapunov: An equilibrium x^∞ is *stable* if for every $\epsilon_1 > 0$, there exists $\epsilon_2 > 0$ with

$$x(0) \in \mathbb{B}(x^\infty; \epsilon_2) \implies x(t) \in \mathbb{B}(x^\infty; \epsilon_1) \quad \forall t \in [0, \infty). \tag{3.3}$$

Thus, the red equilibrium pictured in Figure 3.6 is stable, while the blue equilibrium is not.

A closer look at Figure 3.6 reveals that the trajectories starting in the red circle not only stay within that circle but actually evolve to the equilibrium at (0.45, 1.2). An equilibrium x^∞ is called *asymptotically stable* if it is stable and has the following additional property: there exists $\epsilon > 0$ with

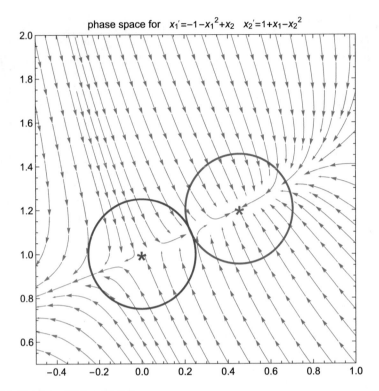

Fig. 3.6 Stable equilibrium (in red)

$$x(0) \in \mathbb{B}(x^{\infty}; \epsilon) \quad \Longrightarrow \quad x(0) \in \text{Basin}(x^{\infty}). \tag{3.4}$$

In the next chapter, we will develop and analyze analogues of stability and asymptotic stability for the attractors associated with the numerical minimization of an objective function.

3.3 Equilibria as Minimizers

There is a fundamental connection between minimization of an objective function $f : \mathbb{R}^n \to \mathbb{R} \cup \{\infty\}$ and the *subgradient system*:

$$x' \in -\partial f(x), \tag{3.5}$$

where $\partial f(x)$ denotes the set of subgradients of f at x. Here, we follow [21, Definition 8.3] in defining subgradients as the following outer limit:

$$\partial f(x) := \{v \in \mathbb{R}^n : \exists\, v^k \to v \text{ and } x^k \to x \text{ with } f(x^k) \to f(x) \text{ and } v^k \in \hat{\partial} f(x^k)\},$$

where $v^k \in \hat{\partial} f(x^k)$ means

$$\liminf_{x \to x^k, \, x \neq x^k} \frac{f(x) - f(x^k) + \langle v^k, x - x^k \rangle}{\|x - x^k\|} \geq 0.$$

(Note that this lim inf is taken with respect to all sequences of $x \neq x^k$ converging to the fixed vector x^k.) When f is smooth (i.e., continuously differentiable), the usual gradient $\nabla f(x)$ is the lone subgradient $\partial f(x) = \{\nabla f(x)\}$, and (3.5) becomes the *gradient system* $x' = -\nabla f(x)$. Notice that the system (3.1) is a gradient system for the objective function $f(x_1, x_2) = x_1 + \frac{x_1^3}{3} - x_1 x_2 - x_2 + \frac{x_2^3}{3}$ since

$$x_1' = -1 - x_1^2 + x_2 = -f_{x_1}(x_1, x_2)$$
$$x_2' = 1 + x_1 - x_2^2 = -f_{x_2}(x_1, x_2).$$

When $f : \mathbb{R}^n \to \mathbb{R} \cup \{\infty\}$ is *proper* (i.e., $f(x) < \infty$ for some $x \in \mathbb{R}$), the generalized Fermat rule $0 \in \partial f(x^\infty)$ is necessary for x^∞ to be a local minimizer of f [21, Theorem 10.1]. It follows immediately that every local minimizer x^∞ of a proper objective function f is an equilibrium of the corresponding subgradient system (3.5). When f is also convex, the implication goes both ways: x^∞ is an equilibrium of the subgradient system (3.5) if and only if x^∞ is a (global) minimizer of f; because in the convex case, the generalized Fermat rule $0 \in \partial f(x^\infty)$ is necessary and sufficient for x^∞ to be a global minimizer of f.

The trajectories in gradient systems $x' = -\nabla f(x)$ evidently follow the steepest descent direction of the objective function f, so it is not surprising that f is monotone (non-increasing) along any state $x(t)$ satisfying the gradient system. This follows immediately from the simple chain rule calculation:

$$\frac{d}{dt} f(x(t)) = \langle \nabla f(x(t)), x'(t) \rangle = -\|\nabla f(x(t))\|^2 \leq 0.$$

This property helps to show that the local minimizers of an analytic objective function f are the same as the asymptotically stable equilibria of the corresponding gradient system [2]. In the next chapter, we will develop analogous results connecting minimization of a lower-semicontinuous objective function f with the asymptotic stability of the attractors associated with the numerical minimization of f, and these results will rely on an important monotonicity property we will call "minvalue-monotonicity."

Chapter 4
Attractors

Because iteration mappings are associated with an objective function $f : \mathbb{R}^n \to \mathbb{R} \cup \{\infty\}$, it will be useful to define the multiset mapping $\operatorname{argmin}_f : \mathbb{R}^n \rightrightarrows \mathbb{R}^n$ which returns all elements of an iterate-multiset X that achieve the lowest f-value:

$$\operatorname*{argmin}_f(X) := \operatorname*{argmin}_{x \in X} f(x) = \left\{ x' \in_\delta X : f(x') = \min_{x \in X} f(x) \right\}.$$

We use this to define the *attractor mapping* $A_{I_{\mathcal{M},f}} : \mathbb{R}^n \rightrightarrows \mathbb{R}^n$ by:

$$A_{I_{\mathcal{M},f}}(X) := \limsup_{k \to \infty} \operatorname*{argmin}_f \left(I^{(k)}_{\mathcal{M},f}(X) \right),$$

and we call any element $x^\infty \in \operatorname{rge}\left(A_{I_{\mathcal{M},f}}\right)$ in the range of the attractor mapping $A_{I_{\mathcal{M},f}}$ an *attractor* for $I_{\mathcal{M},f}$. These objects are analogues of equilibria in dynamical systems, with the important distinction that an attractor depends both on the minimization method $\mathcal{M}$ and on the objective function f. As a consequence, an element x^∞ could be an attractor when minimizing f via one method, but not when minimizing the same objective function f via another method.

Remark 4.1 The outer limit defining the attractor mapping $A_{I_{\mathcal{M},f}}$ means that attractors may be cluster points, and not necessarily limit points of elements in the iterate-multisets. This allows the analysis of examples like the one we saw in the first part of Remark 2.6.

© The Author(s), under exclusive licence to Springer Nature Switzerland AG 2018
A. B. Levy, *Attraction in Numerical Minimization*, SpringerBriefs in Optimization,
https://doi.org/10.1007/978-3-030-04049-9_4

4.1 Local Dense Viability

One of our results relies on the set of all viable initial iterate-multisets $\mathcal{V}_{\mathcal{M},f}$ having the following *local dense viability property near* x^∞: there exists $\epsilon_1 > 0$ such that for every $x \in \mathbb{B}(x^\infty; \epsilon_1)$ and every $\epsilon_2 > 0$, there is a viable iterate-multiset $X \in \mathcal{V}_{\mathcal{M},f}$ with $x \in X$ and $\text{Filter}(X) \subseteq \mathbb{B}(x; \epsilon_2)$.

The set $\mathcal{V}_{\text{NM},f}$ satisfies the local dense viability property near any $x^\infty \in \mathbb{R}^n$ with arbitrarily large ϵ_1 since any point $x \in \mathbb{R}^n$ can be a vertex in a non-degenerate n-simplex whose other vertices are arbitrarily close to x. In the special case when the iterate-multisets have cardinality $m = 1$, the local dense viability property near x^∞ simply requires that there exists $\epsilon_1 > 0$ with $\mathbb{B}(x^\infty; \epsilon_1) \subseteq \mathcal{V}_{\mathcal{M},f}$.

Local dense viability provides a guide for how we should construct iterate-multisets. For instance, we should not include the trial points $\{x^k \pm e_i : i = 1, \ldots, n\}$ in iterate-multisets associated with the coordinate-search method; since otherwise there would be no viable iterate-multisets X satisfying $\text{Filter}(X) \subseteq \mathbb{B}(x^k; \epsilon_2)$ for small ϵ_2, because $x^k \pm e_i \notin \mathbb{B}(x^k; \epsilon_2)$ when $\epsilon_2 < 1$.

When the cardinality $m = |X^k|$ of the iterate-multisets satisfies $m \geq 2$, we will also use the following *restricted local dense viability property near* x^∞: there exists $\epsilon_1 > 0$ such that for every $x \in \mathbb{B}(x^\infty; \epsilon_1) \setminus \{x^\infty\}$ and every $\epsilon_2 > 0$, there is a viable iterate-multiset $X \in \mathcal{V}_{\mathcal{M},f}$ with $x, x^\infty \in X$ and $\text{Filter}(X) \setminus \{x^\infty\} \subseteq \mathbb{B}(x; \epsilon_2)$. Notice that $m \geq 2$ is necessary since otherwise we cannot have both $x \neq x^\infty$ and x^∞ as elements of X.

The set $\mathcal{V}_{\text{NM},f}$ also satisfies the restricted local dense viability property near any $x^\infty \in \mathbb{R}^n$ with arbitrarily large ϵ_1, since any two distinct points x^∞ and x in $\mathbb{R}^n$ can be supplemented by $n - 1$ other points, all arbitrarily close to x, to form a non-degenerate n-simplex.

4.2 Basins of Attraction

We define the *basin of attraction* $\text{Basin}_{\mathcal{M},f}(x^\infty)$ associated with an attractor x^∞ for $I_{\mathcal{M},f}$ to be the collection of all multisets X for which x^∞ is in the image of the attractor mapping:

$$\underset{\mathcal{M},f}{\text{Basin}}(x^\infty) := \left\{ X : x^\infty \in A_{I_{\mathcal{M},f}}(X) \right\}.$$

It is sometimes useful to extend this notion to any set S of attractors, in which case we get the following collection of multisets:

$$\underset{\mathcal{M},f}{\text{Basin}}(S) := \left\{ X : S \cap A_{I_{\mathcal{M},f}}(X) \neq \emptyset \right\},$$

attracted to at least one of the attractors x^∞ in S.

Remark 4.2 Unlike the situation in dynamical systems, basins of attraction associated with different attractors can overlap. For instance, the first example in
Remark 2.6 has iteration mapping $I_{SD_1, x_1^2 + x_2^2}(x) = -x$, and every point $x^\infty \in$
$\mathbb{R}^2 \setminus (0, 0)$ is an attractor with basin of attraction:

$$\underset{\mathscr{M}, f}{\text{Basin}}(x^\infty) = \{x^\infty, -x^\infty\} = \underset{\mathscr{M}, f}{\text{Basin}}(-x^\infty).$$

4.3 Stability

Our first notion of stability for an attractor is an analogue of stability in dynamical
systems (3.3). We say that an attractor x^∞ for $I_{\mathscr{M}, f}$ is *stable* if for every $\epsilon_1 > 0$,
there exists $\epsilon_2 > 0$ with:

$$\text{Filter}(X) \subseteq \mathbb{B}(x^\infty; \epsilon_2) \text{ and } X \in \mathscr{V}_{\mathscr{M}, f}$$

$$\implies \text{Filter}\left(I^{(k)}_{\mathscr{M}, f}(X)\right) \subseteq \mathbb{B}(x^\infty; \epsilon_1) \quad \forall k \geq 1.$$

As is the case with the attractors themselves, stability depends on the minimization
method $\mathscr{M}$ as well as on the objective function f; so, an attractor x^∞ could exhibit
stability for one method but not for another.

Recall that in dynamical systems, asymptotic stability is stability with an
additional property. To develop our analogue of that additional property, we define
the *radius of attraction* to x^∞:

$$\underset{\mathscr{M}, f}{\text{Radius}}(x^\infty) := \sup\{\epsilon > 0 \mid \text{Filter}(X) \subseteq \mathbb{B}(x^\infty; \epsilon) \text{ and } X \in \mathscr{V}_{\mathscr{M}, f}$$

$$\implies X \in \underset{\mathscr{M}, f}{\text{Basin}}(x^\infty)\},$$

which quantifies the relative intensity of attraction to x^∞. The radius of attraction
always satisfies $\text{Radius}_{\mathscr{M}, f}(x^\infty) \geq 0$, and is positive (including ∞) if and only if
there exists an $\epsilon > 0$ such that x^∞ attracts every viable initial iterate-multiset whose
elements are within ϵ of x^∞. A stable attractor x^∞ is *asymptotically stable* if it has
positive radius of attraction.

Remark 4.3 The requirement of viability $X \in \mathscr{V}_{\mathscr{M}, f}$ in these definitions is crucial.
For instance in the Nelder–Mead method, there are degenerate simplices X whose
elements are all arbitrarily close to an arbitrary point $x^\infty \in \mathbb{R}^n$. These X are not
in the domain of $I_{NM, f}$, and hence not in $\text{Basin}_{NM, f}(x^\infty)$. It follows that if we
did not require viability in our definitions, every attractor x^∞ for $I_{NM, f}$ would be
stable (trivially, since $\emptyset \subseteq \mathbb{B}(x^\infty; \epsilon_1)$), and no attractor x^∞ for $I_{NM, f}$ would be
asymptotically stable.

The next corollary is an immediate consequence of Theorem 2.1.

Corollary 4.1 *An attractor x^∞ for $I_{\mathcal{M},f}$ has positive radius of attraction as long as: (i) $I_{\mathcal{M},f}$ is $\mathbf{d}_{\mathrm{exc}}$-contractive at X^∞, and (ii) the iterate-multiset $X^\infty :=$ $\{x^\infty, \ldots, x^\infty\}$ is fixed by $I_{\mathcal{M},f}$. Moreover, condition (i) is guaranteed in this case when the upper $\mathbf{d}_{\mathrm{exc}}$-derivative satisfies $D_{\mathbf{d}_{\mathrm{exc}}}^+ I_{\mathcal{M},f}(X^\infty) < 1$.*

Proof Only the final statement needs justification, and it follows from Corollary 1.2 since the iteration mapping $I_{\mathcal{M},f}$ in this case satisfies

$$\mathbf{d}_{\mathrm{exc}}(X, X^\infty) = 0 \implies X = X^\infty \implies I_{\mathcal{M},f}(X) = X^\infty$$
$$\implies \mathbf{d}_{\mathrm{exc}}\left(I_{\mathcal{M},f}(X), X^\infty\right) = 0$$

when $X \in \mathrm{dom}\left(I_{\mathcal{M},f}\right)$.   □

Remark 4.4 This corollary holds with $\mathbf{d}_{\mathrm{sup}}$ in place of $\mathbf{d}_{\mathrm{exc}}$.

Remark 4.5 The assumptions of Corollary 4.1 actually guarantee the stronger result that the attractor is the limit of every element x_i^k in the iterate-multisets X^k generated by nearby viable initial iterate-multisets.

We can weaken the assumptions in Corollary 4.1 to achieve the same conclusion by appealing instead to Theorem 2.2.

Corollary 4.2 *An attractor x^∞ for $I_{\mathcal{M},f}$ has positive radius of attraction as long as: (i) $I_{\mathcal{M},f}$ is $\mathbf{d}_{\mathrm{exc}}$-non-expansive at X^∞ with radius $\delta > 0$ such that*

$$\sup_{X^0 \in \mathcal{V}_{\mathcal{M},f} \cap \left\{X : \mathbf{d}_{\mathrm{exc}}\left(X, X^\infty\right) \leq \delta\right\}} \liminf_{k \to \infty} \frac{\mathbf{d}_{\mathrm{exc}}\left(I_{\mathcal{M},f}^{(k+1)}(X^0), X^\infty\right)}{\mathbf{d}_{\mathrm{exc}}\left(I_{\mathcal{M},f}^{(k)}(X^0), X^\infty\right)} < 1 \qquad (4.1)$$

holds, and (ii) the iterate-multiset $X^\infty := \{x^\infty, \ldots, x^\infty\}$ is fixed by $I_{\mathcal{M},f}$.

Proof This follows immediately from Theorem 2.2 since $I_{\mathcal{M},f}^{(k)}(X^0) = X^k$ and $I_{\mathcal{M},f}^{(k+1)}(X^0) = I_{\mathcal{M},f}(X^k)$, so the bound (4.1) ensures that (2.1) holds for all viable initial iterate-multisets X^0 satisfying $\mathbf{d}_{\mathrm{exc}}\left(X^0, X^\infty\right) \leq \delta$.   □

Remark 4.6 This corollary holds with $\mathbf{d}_{\mathrm{sup}}$ in place of $\mathbf{d}_{\mathrm{exc}}$.

Even when an attractor x^∞ has radius of attraction equal to zero, it can still have a positive *radius of restricted attraction* $\mathrm{RadRes}_{\mathcal{M},f}(x^\infty)$, where we add $x^\infty \in X$ to the conditions in the definition of $\mathrm{Radius}_{\mathcal{M},f}(x^\infty)$:

$$\mathrm{RadRes}_{\mathcal{M},f}(x^\infty) := \sup\{\epsilon > 0 | \mathrm{Filter}(X) \subseteq \mathbb{B}(x^\infty; \epsilon), \ X \in \mathcal{V}_{\mathcal{M},f}, \text{ and } x^\infty \in X$$
$$\implies X \in \mathrm{Basin}_{\mathcal{M},f}(x^\infty)\}.$$

The radius of restricted attraction is really only interesting when the iterate-multisets have cardinality $m \geq 2$, since when $m = 1$ it reduces to:

$$\underset{\mathscr{M},f}{\mathrm{RadRes}}(x^\infty) = \begin{cases} \infty \text{ if } x^\infty \in \underset{\mathscr{M},f}{\mathrm{Basin}}(x^\infty) \cap \mathscr{V}_{\mathscr{M},f} \\ 0 \text{ otherwise.} \end{cases}$$

A stable attractor x^∞ is *restricted asymptotically stable* if it has positive radius of restricted attraction. Notice that the construction of the radius of restricted attraction guarantees that it is always at least as large as the radius of attraction:

$$\underset{\mathscr{M},f}{\mathrm{RadRes}}(x^\infty) \geq \underset{\mathscr{M},f}{\mathrm{Radius}}(x^\infty).$$

It follows immediately that an asymptotically stable attractor is always restricted asymptotically stable.

4.4 Minvalue-Monotonicity

We say that an iteration mapping $I_{\mathscr{M},f}$ is *minvalue-monotone* if for every viable initial iterate-multiset $X^0 \in \mathscr{V}_{\mathscr{M},f}$, the minimum value:

$$m^k := \min_{x \in I_{\mathscr{M},f}^{(k)}(X^0)} f(x)$$

never increases from one iteration to the next: $m^k \leq m^{k-1}$ for all $k \geq 1$. This is a key property for connecting attraction to minimality, and it holds for all of our example iteration mappings (regardless of objective function). In fact, three of our example mappings, coordinate-search $I_{\mathrm{CS},f}$ and the final two steepest-descent mappings $I_{\mathrm{SD3},f}$ and $I_{\mathrm{SD4},f}$, have the following even stronger property: an iteration mapping $I_{\mathscr{M},f}$ is *strictly minvalue-monotone* if for every viable initial iterate-multiset $X^0 \in \mathscr{V}_{\mathscr{M},f}$, the minimum value decreases whenever the iterate-multiset changes: $m^k < m^{k-1}$ for $k \geq 1$ with $I_{\mathscr{M},f}^{(k)}(X^0) \neq I_{\mathscr{M},f}^{(k-1)}(X^0)$.

Remark 4.7 The classical monotone convergence theorem guarantees that minvalue-monotonicity (or strict minvalue-monotonicity) implies that the sequence of minimum values $\{m^k\}$ converges as long as the objective function f is bounded below on the collection of iterate-multisets X^k. Note that this does not necessarily mean that we can conclude something similar about the iterate-multisets X^k themselves.

4.5 Attractors as Minimizers

Our next result gives conditions under which an attractor is guaranteed to be a local minimizer of the objective function.

Theorem 4.1 *Assume that f is lower-semicontinuous at an attractor x^∞ for the minvalue-monotone iteration mapping $I_{\mathcal{M},f}$ and that the local dense viability property holds near x^∞. If its radius of attraction is positive, then x^∞ is a local minimizer of f.*

Proof Define $\epsilon := \min\{\mathrm{Radius}_{\mathcal{M},f}(x^\infty), \epsilon_1\}$, where $\epsilon_1 > 0$ comes from the local dense viability property near x^∞. Choose any point $x \in \mathbb{B}\left(x^\infty; \frac{\epsilon}{3}\right)$ and apply the dense viability property with $\epsilon_2 := \frac{\epsilon}{3}$ to find a viable initial iterate-multiset X^0 with $x \in X^0$ and $\mathrm{Filter}(X^0) \in \mathbb{B}\left(x; \frac{\epsilon}{3}\right)$. It follows that $\mathrm{Filter}(X^0) \in \mathbb{B}\left(x^\infty; \frac{2}{3}\epsilon\right)$, and since $\frac{2}{3}\epsilon < \mathrm{Radius}_{\mathcal{M},f}(x^\infty)$, we conclude that $X^0 \in \mathrm{Basin}_{\mathcal{M},f}(x^\infty)$. Thus, there exist $N \in \mathcal{N}_\infty^{\#}$ and $x^k \in \mathrm{argmin}_f\left(I_{\mathcal{M},f}^{(k)}(X^0)\right)$ for $k \in N$ with $x^k \xrightarrow{N} x^\infty$. Since $x \in X^0$ and the iteration mapping $\mathscr{I}_{\mathcal{M},f}$ is minvalue-monotone, it follows that

$$f(x^k) \le f(x) \quad \text{for all } k \in N.$$

We then conclude from the lower semicontinuity of f at x^∞ that

$$f(x^\infty) \le \liminf_{k \in N} f(x^k) \le f(x).$$

Since $x \in \mathbb{B}\left(x^\infty; \frac{\epsilon}{3}\right)$ is arbitrary, this means that $f(x^\infty)$ does not exceed the f-value of any point in $\mathbb{B}\left(x^\infty; \frac{\epsilon}{3}\right)$. Hence, x^∞ is a local minimizer of f as claimed.
□

Remark 4.8 Lower semicontinuity at x^∞ is needed to rule out functions like

$$f(x) = \begin{cases} 1 & \text{if } x = 0 \\ x^2 & \text{otherwise} \end{cases}$$

(from [1]).

Remark 4.9 The implication in Theorem 4.1 cannot be reversed in general, as we will see in the following example:

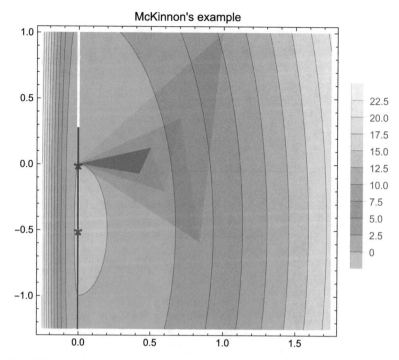

Fig. 4.1 Nelder–Mead attracted to a non-minimizer along a segment (in blue) of attractors

McKinnon's Example

McKinnon [17] constructed the family of convex continuous functions on $\mathbb{R}^2$:

$$f(x_1, x_2) := \begin{cases} \theta \phi |x_1|^\tau + x_2 + x_2^2 & \text{when } x_1 \leq 0 \\ \theta x_1^\tau + x_2 + x_2^2 & \text{otherwise} \end{cases} \quad (4.2)$$

(for $\tau > 0$, $\theta > 0$, and $\phi > 0$) with a single global minimizer at $(0, -0.5)$ (shown as a red star on the contour plot in Figure 4.1 of McKinnon's function with $\tau = 2$, $\theta = 6$, and $\phi = 60$).

McKinnon's functions with $\tau > 1$ are smooth and strictly convex, which we would expect to be sufficient structure to support a traditional result like the convergence of iterate-multisets to the minimizer. However, [17] gives

(continued)

conditions under which the Nelder–Mead method will generate the constant sequence of best vertices $(0, x_2^\infty) = \mathrm{argmin}_f(X^k)$ when the iterate-multisets X^k are initiated at:

$$X^0 = \left\{ (0, x_2^\infty), (v_1^{p+1}, x_2^\infty + v_2^{p+1}), (v_1^p, x_2^\infty + v_2^p) \right\}, \tag{4.3}$$

for $v_1 = \frac{1+\sqrt{33}}{8}$, $v_2 = \frac{1-\sqrt{33}}{8}$, and any $p \in \mathbb{N}$. For instance, when $\tau = 2$, $\theta = 6$, and $\phi = 60$ in McKinnon's function (4.2), these necessary conditions are satisfied as long as $x_2^\infty \in (-1.24, 0.27)$. Figure 4.1 shows this interval as a blue segment along the x_2-axis, and the largest purple triangle represents an initial iterate-multiset of the form (4.3) with $x_2^\infty = 0$. The smaller (nested) purple triangles represent the subsequent iterate-multisets generated by the Nelder–Mead method, and notice that the best vertex never changes in this case from $(0, x_2^\infty)$.

Since $p \in \mathbb{N}$ can be arbitrarily large, and $v_1, v_2 \in (-1, 1)$, we know that for any fixed $x_2^\infty \in (-1.24, 0.27)$ and any $\epsilon > 0$ we can find an open interval O around x_2^∞ such that for every $\hat{x}_2^\infty \in O \setminus \{x_2^\infty\}$, there is a (viable) initial iterate-multiset $X^0 \in \mathbb{B}^m \left((0, x_2^\infty); \epsilon \right)$ of the form (4.3), for which the consequent Nelder–Mead best vertices are always $(0, \hat{x}_2^\infty) \neq (0, x_2^\infty)$. This means in particular that none of the attractors $(0, x_2^\infty)$ has positive radius of attraction. Of course, this is guaranteed by (the contrapositive of) Theorem 4.1 for all but the minimizer at $(0, -0.5)$ since there are no other local minimizers for this function. The fact that the minimizer at $(0, -0.5)$ in this case also has zero radius of attraction means that the implication in Theorem 4.1 cannot be reversed in general.

We have just seen that the minimizer of McKinnon's function (4.2) has zero radius of attraction; however, simulations suggest that it has positive radius of restricted attraction. If we change the local dense viability assumption (to restricted local dense viability) in Theorem 4.1, we get the following analogous result when x^∞ has positive radius of restricted attraction.

Theorem 4.2 *Assume that f is lower-semicontinuous at an attractor x^∞ for the iteration mapping $I_{\mathcal{M}, f}$, that $I_{\mathcal{M}, f}$ is minvalue-monotone, and that the restricted local dense viability property holds near x^∞. If its radius of restricted attraction is positive, then x^∞ is a local minimizer of f.*

Proof This follows in a very similar manner to the proof of Theorem 4.1, but now we define $\epsilon := \min\{\mathrm{RadRes}_{\mathcal{M}, f}(x^\infty), \epsilon_1\}$, where $\epsilon_1 > 0$ comes from the restricted local dense viability property near x^∞. Then, we choose any $x \in \mathbb{B}\left(x^\infty; \frac{\epsilon}{3}\right) \setminus \{x^\infty\}$

and apply the restricted local dense viability property with $\epsilon_2 := \frac{\epsilon}{3}$ to find a viable initial iterate-multiset X^0 with x, $x^\infty \in X^0$ and Filter$(X^0) \setminus \{x^\infty\} \in \mathbb{B}\left(x; \frac{\epsilon}{3}\right)$. It follows that Filter$(X^0) \in \mathbb{B}\left(x^\infty; \frac{2}{3}\epsilon\right)$, and since $\frac{2}{3}\epsilon < \mathrm{RadRes}_{\mathcal{M},f}(x^\infty)$, we conclude that $X^0 \in \mathrm{Basin}_{\mathcal{M},f}(x^\infty)$. Then, the same argument as in Theorem 4.1 allows us to deduce that $f(x^\infty)$ does not exceed the f-value of any point in $\mathbb{B}\left(x^\infty; \frac{\epsilon}{3}\right)$. $\qquad\square$

Since simulations suggest that the minimizer of McKinnon's function (4.2) has positive radius of restricted attraction, we suspect that we can prove implications in the opposite direction to Theorem 4.2. We now prove that this is the case whenever an attractor is also stable.

Theorem 4.3 *A stable attractor x^∞ for $I_{\mathcal{M},f}$ has positive radius of restricted attraction (and hence is restricted asymptotically stable) under either of the following pairs of conditions:*

i. $I_{\mathcal{M},f}$ is strictly minvalue-monotone and x^∞ is a local minimizer of f, or
ii. $I_{\mathcal{M},f}$ is minvalue-monotone and x^∞ is a strict local minimizer of f.

Proof Since x^∞ is stable, there exists $\epsilon_2 > 0$ such that any viable initial iterate-multiset X^0 satisfying Filter$(X^0) \subseteq \mathbb{B}(x^\infty; \epsilon_2)$ will generate iterate-multisets $X^k = I_{\mathcal{M},f}^{(k)}(X^0)$ that always stay within the neighborhood where x^∞ is a local minimizer (strict local minimizer). Then, the assumption that $I_{\mathcal{M},f}$ is strictly minvalue-monotone (minvalue-monotone) ensures that $x^\infty \in X^k$ for all $k \geq 1$ as long as $x^\infty \in X^0$. In this case, we conclude that $x^\infty \in A_{I_{\mathcal{M},f}}(X^0)$, so that $\mathrm{RadRes}_{\mathcal{M},f}(x^\infty) \geq \epsilon_2$. $\qquad\square$

Remark 4.10 Condition (ii) of this theorem applies to our example of the strict global minimizer $x^\infty = (0, -0.5)$ of McKinnon's function (4.2) with parameters $\tau = 2$, $\theta = 6$, and $\phi = 60$, since the iteration mapping $I_{\mathrm{NM},f}$ in this case is minvalue-monotone. Moreover, the minimizer at x^∞ is a stable attractor, which follows in this case since the objective function is strictly convex and has bounded level-sets; so for any $\epsilon_1 > 0$, we can always choose $\epsilon_2 > 0$ small enough so that Filter$(X^0) \subseteq \mathbb{B}((0, -0.5); \epsilon_2)$ ensures that the level-set $\mathrm{lev}_0(f)$ is contained in $\mathbb{B}((0, -0.5); \epsilon_1)$ where:

$$\mathrm{lev}_k(f) := \left\{ x \in \mathbb{R}^n : f(x) \leq \max_{x' \in X^k} f(x') \right\}.$$

From [15, Lemma 3.5], we have that the Nelder–Mead method will never use shrink steps in this case (due to the strict convexity), which ensures that we have what [16] calls "f-stability": Filter$\left(X^{k+1}\right) \subseteq \mathrm{lev}_k(f)$ for all $k \geq 1$. It follows that $\mathrm{lev}_k(f) \subseteq \mathrm{lev}_0(f)$ for all $k \geq 1$, so we conclude stability from the fact that $\mathrm{lev}_0(f) \subseteq \mathbb{B}((0, -0.5); \epsilon_1)$. Thus, Theorem 4.3 confirms that x^∞ has positive radius of restricted attraction, as our simulations suggested.

Remark 4.11 Recall that our notions of attraction and various kinds of stability depend both on the objective function f and on the minimization method $\mathcal{M}$. Thus for a given objective f, the conditions of the preceding theorems may be met by some method $\mathcal{M}$ even if they are not met for other methods. This highlights a potential benefit in practice of choosing a minimization method to suit a particular objective function.

Chapter 5
Basin Analysis via Simulation

Recall from Section 4.2 that the basin of attraction $\text{Basin}_{\mathcal{M},f}(S)$ associated with any set S of attractors is the collection of multisets attracted to at least one of the attractors x^∞ in S. In general, the most practical way to estimate basins of attraction is through simulation; and it can be relatively easy to illustrate simulated basins when the objective f is a function of two variables, so we'll focus here on examples of this type.

In order to easily illustrate our simulations, we will explore the steepest-descent method (with three different line-searches) and the coordinate-search method since these all generate iterates x^k (i.e., singleton iterate-multisets). Our simulations use 10,000 initial iterates chosen pseudo-randomly from a square section of the $x_1 x_2$-plane, and color-coded to match their corresponding attractor. In a separate subsection on the Nelder–Mead method, we will discuss a few different options for illustrating simulated basins associated with methods (like Nelder–Mead) whose iterate-multisets are not singleton.

We also classify basin size via an approach motivated by the one for dynamical systems [22] that we discussed in Chapter 3. Unlike [22], our results depend on the pre-distance function $\mathbf{d}$ we use to measure how far an initial iterate-multiset X^0 is from the center of mass $\bar{x}^\infty$ of the set of attractors S in $\mathbb{R}^n$. We signal this dependence with a subscript "$\mathbf{d}$" in our model:

$$P_{\mathbf{d}}(r) = P_0\, r^{-\gamma} \qquad (5.1)$$

for the asymptotic behavior of the probability that an initial iterate-multiset X^0 satisfying $\mathbf{d}\left(X^0, \bar{x}^\infty\right) \leq r$ also satisfies $X^0 \in \text{Basin}_{\mathcal{M},f}(S)$. In one case, we will refer to the *effective radius* which is the value of r at which the asymptotic model (5.1) returns probability 1. The main size-classes are as follows (from largest to smallest):

© The Author(s), under exclusive licence to Springer Nature Switzerland AG 2018
A. B. Levy, *Attraction in Numerical Minimization*, SpringerBriefs in Optimization,
https://doi.org/10.1007/978-3-030-04049-9_5

Classes of Basin Size

1. $P_0 > 0$ and $\gamma = 0$; $\text{Basin}_{\mathcal{M},f}(S)$ contains approximately P_0 of all iterate-multisets.
2. $P_0 > 0$ and $\gamma \in (0, n]$; $\text{Basin}_{\mathcal{M},f}(S)$ has effective radius $r_0 := P_0^{\frac{1}{\gamma}}$.
3. $P_0 = 0$; $\text{Basin}_{\mathcal{M},f}(S)$ contains no pseudo-randomly generated iterate-multisets.

Note that we have made the following modifications to the classifications from [22]: we consolidated analogues of the first and second classes from [22] into a single first class here, we consolidated analogues of the third and fourth classes from [22] into a single second class, and we added a new fourth class for the case when $P_{\mathbf{d}}(r) = 0$.

Our basin-sizing procedure begins by pseudo-randomly choosing m elements from $\mathbb{R}^n$ in such a way that the initial iterate-multiset X^0 consisting of those m elements satisfies $\mathbf{d}\left(X^0, \bar{x}^\infty\right) \leq 1$. We generate 1000 samples X^0 in this manner and assign $P_{\mathbf{d}}(1)$ to be the fraction of them in $\text{Basin}_{\mathcal{M},f}(S)$. Then, we determine estimates of $P_{\mathbf{d}}(2^d)$ via the recursive formula:

$$P_{\mathbf{d}}\left(2^d\right) = 2^{-n}\, P_{\mathbf{d}}\left(2^{d-1}\right) + \left(1 - 2^{-n}\right) \Delta P_{\mathbf{d}}\left(2^{d-1}\right) \qquad \text{for } d = 1, \ldots, 10;$$
(5.2)

where each $\Delta P_{\mathbf{d}}\left(2^{d-1}\right)$ is estimated by testing the basin membership of 1000 initial iterate-multisets generated in the same way as above, but now with m elements pseudo-randomly chosen so that the resulting X^0 satisfies $2^{d-1} < \mathbf{d}\left(X^0, \bar{x}^\infty\right) \leq 2^d$. Finally, we fit a line $a\,d + b$ to the final four points on the $\log_2$-$\log_2$ plot of our 2^d versus $P\left(2^d\right)$ data (excluding data points with $P\left(2^d\right) = 0$ since the $\log_2$ of these is undefined) to determine the values of $\gamma = -a$ and $P_0 = 2^b$ in the asymptotic model $P_{\mathbf{d}}(r) = P_0\, r^{-\gamma}$. When the iterate-multisets have cardinality $m = 1$, every pre-distance function reduces to $\mathbf{d}(x^0, \bar{x}^\infty) = \|x^0 - \bar{x}^\infty\|$, so we drop the subscript "$\mathbf{d}$" from the notation.

Finally, we modify the approach from [9] for dynamical systems in order to calculate "basin entropy" for each basin of attraction, which quantifies the cumulative unpredictability of basin membership for initial iterate-multisets whose elements all come from a bounded region Ω of the domain space $\mathbb{R}^n$ for the objective function. To do this, we consider the region:

$$\Omega^m = \underbrace{\Omega \times \Omega \times \ldots \times \Omega}_{m\text{ times}}$$

consisting of m-copies of Ω in the space $\mathbb{R}^{mn}$, where m is the cardinality of the iterate-multisets as usual. We subdivide $\Omega^m \subseteq \mathbb{R}^{mn}$ into a grid of N equal boxes (mn-hypercubes with edge-length ϵ) and compute the probability $p_{i,j}$ of each basin membership profile j (out of the J_i possible profiles in box i) as the fraction of 25 different runs of the minimization method initiated pseudo-randomly in box i that have profile j. We define the *basin entropy Sb* to be the average Gibbs entropy over all N boxes:

$$Sb := \frac{1}{N} \sum_{i=1}^{N} \sum_{j=1}^{J_i} \frac{p_{i,j}}{\log(J_i)} \log\left(\frac{1}{p_{i,j}}\right). \tag{5.3}$$

Note that the basin entropy generally changes with a different choice of N, and should approach zero as N approaches infinity. Note also that we have normalized Sb (dividing by $\log(J_i)$) so that it always satisfies $0 \leq Sb \leq 1$.

In our first four examples (where $n = 2$ and $m = 1$), we illustrate our entropy results with a probability plot where boxes are colored on a hue-scale between red and blue according to the probability of initial iterate-multisets generated from that box are in the basin of attraction for the red or blue attractor set. The highest-entropy boxes are colored purple and are found at the boundaries between basins of attraction. In our first four examples, all of the initial iterates are either attracted to one attractor or the other, in which case the J_i are equal to 2. In the section on the Nelder–Mead method, we use a nested probability plot which we describe in detail there.

5.1 Example 1: Two Global Minimizers

The objective function:

$$f_1(x_1, x_2) := \left(x_1^2 + (x_2 - 1)^2\right)\left(x_1^2 + (x_2 + 1)^2\right)$$

has two global minimizers, one at $(0, 1)$ and one at $(0, -1)$, as well as a saddle point at $(0, 0)$. The two global minimizers are marked with blue and red stars, respectively, on the contour plot shown in Figure 5.1.

This function is differentiable, so we can examine its associated gradient system:

$$x_1' = -4 x_1 \left(x_1^2 + x_2^2 + 1\right)$$
$$x_2' = -4 x_2 \left(x_1^2 + x_2^2 - 1\right) \tag{5.4}$$

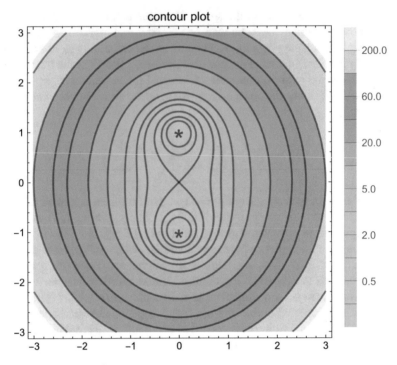

Fig. 5.1 Contour plot of f_1

whose equilibria and basins of attraction are pictured in Figure 5.2. From our discussion in Chapter 3, it should not be surprising that the minimizers in Figure 5.1 appear as equilibria in Figure 5.2.

5.1.1 Simulated Basins

When we apply either the steepest-descent method with the exact line-search SD$_2$ or the no-longer-downhill line-search SD$_3$ to this objective function, we get the exact same simulated basins of attraction, as shown in Figure 5.3. This image happens to also be the same as that which we would get by simulating basins of attraction for the equilibria of the gradient system (5.4), which is not entirely surprising since the trajectories in any gradient system follow the steepest-descent direction of the objective function f. However, we will see in other examples that this is not the case.

If we instead apply steepest-descent with the backtracking line-search SD$_4$ or the coordinate-search method CS, we get the images in Figure 5.4. These basins are clearly different from each other, as well as being different from the basin in Figure 5.3 produced by SD$_2$ or SD$_3$. This highlights the fact that our basins of

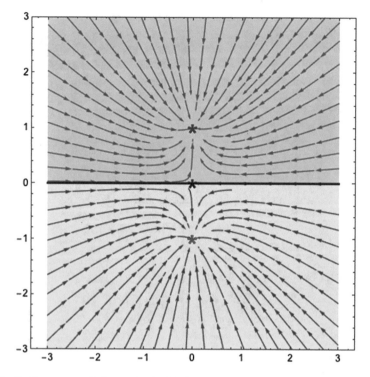

Fig. 5.2 Gradient system (5.4) associated with f_1

attraction depend on the choice of numerical method as well as on the objective function. Notice that for all four minimization methods, 50% of the pseudo-randomly chosen initial iterates are attracted to each global minimizer, which is not surprising given the symmetry exhibited by the objective function here.

It is important to recognize that the images Figures 5.3 and 5.4 are *estimates*, and we should use them with caution to make precise conclusions about the actual basins of attraction. For instance, we can sometimes find additional elements in a basin of attraction if we instead run the method in reverse from pseudo-randomly chosen initial pairs very close to the blue minimizer at $(0, 1)$. In Figure 5.5a, we have superimposed a plot of some larger blue points obtained by reversing SD_4 on the image from Figure 5.4a of our basin simulation for SD_4.

Because of the symmetry in this particular example, we can also generate images related to the basins of attraction by color-coding the initial points that switch to the other side of the x_1-axis after a number of iterations of SD_4. The image in Figure 5.6a shows the initial points in red if their resulting iterate is in the lower-half (where the red minimizer at $(0, -1)$ is located) after twelve iterations of SD_4, and in blue if their resulting iterate is in the upper-half (where the blue minimizer at $(0, 1)$ is located). It is clear from Figures 5.5a and 5.6a that the simulated basin generated by our usual procedure (and shown in Figures 5.5b and 5.6b for comparison) misses

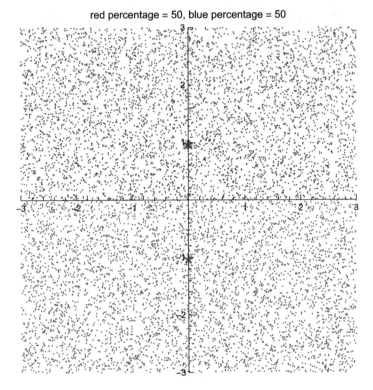

Fig. 5.3 Simulated basins for f_1 with steepest-descent SD_2 or SD_3

a variety of fine details. Of course, we cannot expect our usual simulation procedure to identify isolated points or any other components of basins that have less than full dimension. Nonetheless, the basin estimates that result from simulation are well-aligned with practice where these minimization methods are implemented from pseudo-random initial iterate-multisets. Recall that our notions of basin size and basin entropy also rely on simulation from pseudo-random initial iterate-multisets, and therefore are also well-aligned with the practical implementation of these minimization methods.

5.1.2   Basin Sizes

Using the basin-sizing procedure described at the beginning of this section, we get the estimates in Table 5.1 for the probabilities associated with the basins of attraction for the blue minimizer at $(0, 1)$ and the red minimizer at $(0, -1)$. Notice that the probability estimates of 1.0 for each attractor in the $d = 0$ case for SD_2 and SD_3 in Table 5.1 are consistent with those attractors having a (positive) radius of attraction

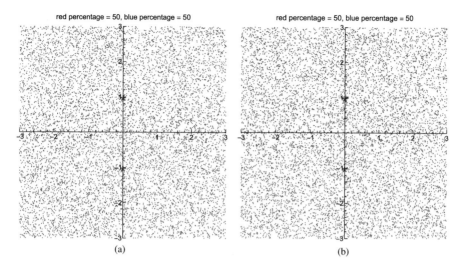

Fig. 5.4 Simulated basins of f_1 with SD$_4$ and CO. (**a**) Steepest-descent SD$_4$. (**b**) Coordinate-search CS

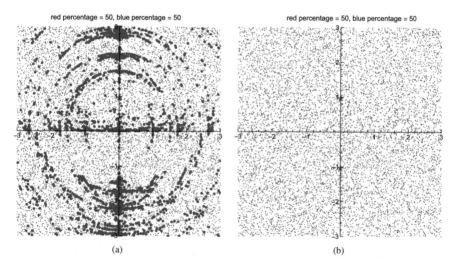

Fig. 5.5 Simulated basins of f_1 with SD$_4$ in reverse and forward. (**a**) Reversing steepest-descent SD$_4$. (**b**) Steepest-descent SD$_4$

at least as large as 1, which is what we would expect in these cases based on the simulated basins shown in Figure 5.3.

We fit a line to the final four points on the $\log_2$ plot of the probability data in Table 5.1 to determine the values of γ and P_0 shown in Table 5.2. Recall from Figures 5.3 and 5.4 that 50% of the pseudo-randomly chosen initial iterates are attracted to $(0, 1)$ and 50% are attracted to $(0, -1)$. Because of the symmetry of this objective function, we would expect the same 50% attraction on any scale. Notice

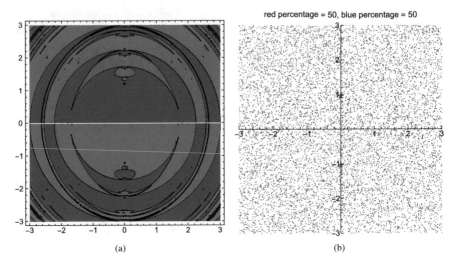

Fig. 5.6 Simulated basins of f_1 with SD$_4$ via side-switching and forward. (**a**) Switching sides after twelve steps of SD$_4$. (**b**) Steepest-descent SD$_4$

Table 5.1 Basin-size data for f_1

		0	1	2	3	4	5	6	7	8	9	10
	d											
SD$_2$	$P(2^d)$	1.0	0.794	0.656	0.589	0.551	0.511	0.511	0.497	0.495	0.503	0.504
	$P(2^d)$	1.0	0.800	0.661	0.580	0.527	0.519	0.525	0.512	0.501	0.504	0.514
SD$_3$	$P(2^d)$	1.0	0.805	0.672	0.580	0.522	0.509	0.508	0.482	0.516	0.482	0.476
	$P(2^d)$	1.0	0.810	0.661	0.570	0.515	0.511	0.504	0.500	0.509	0.506	0.494
SD$_4$	$P(2^d)$	0.820	0.532	0.456	0.486	0.469	0.459	0.474	0.491	0.493	0.500	0.489
	$P(2^d)$	0.842	0.546	0.474	0.470	0.491	0.489	0.489	0.502	0.509	0.505	0.513
CS	$P(2^d)$	0.701	0.688	0.596	0.566	0.535	0.489	0.523	0.521	0.516	0.498	0.493
	$P(2^d)$	0.711	0.695	0.598	0.570	0.517	0.515	0.507	0.506	0.483	0.496	0.497

that in every case shown in Table 5.2, the basins of attraction are in Class 1 so the values of P_0 approximate the attraction percentage. We already know to expect 50% attraction here, so it is not surprising that the P_0 are close to that value. Even more accurate results may be obtained by using a greater number of samples and by generating the data for larger integer values of d.

5.1.3 Basin Entropy

We computed the basin entropy associated with the square region:

$$\Omega = \{(x_1, x_2) \in \mathbb{R}^2 : -3 \leq x_1 \leq 3 \text{ and } -3 \leq x_2 \leq 3\}$$

Table 5.2 Size
classifications for basins of f_1

	γ	P_0	$P(r)$	Class
SD$_2$	0	0.48	0.48	1
	0	0.50	0.50	1
SD$_3$	0	0.53	0.53	1
	0	0.52	0.52	1
SD$_4$	0	0.49	0.49	1
	0	0.48	0.48	1
CS	0	0.60	0.60	1
	0	0.51	0.51	1

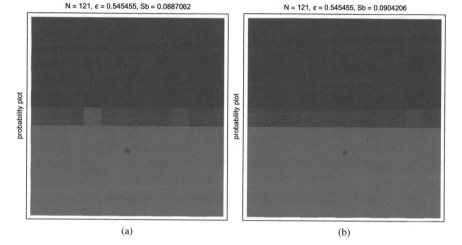

Fig. 5.7 Basin entropy of f_1 with SD$_2$ and SD$_3$. (**a**) Steepest-descent SD$_2$. (**b**) Steepest-descent SD$_3$

and created the corresponding probability plots shown in Figures 5.7 and 5.8. It is not surprising that the two images in Figure 5.7 are so similar, because the simulated basins in these two cases were identical. The higher entropy values for the cases shown in Figure 5.8 are consistent with the more complex basin boundaries we see in Figure 5.4. Notice that in all cases, the probability plots provide a crude form of simulated basin.

5.2   Example 2: One Global Minimizer and a Saddle Point

The objective function:

$$f_2(x_1, x_2) := \frac{1}{2}\left(1 + 2\left(x_1^2 + (x_2 - 1)^2\right)\right)\left(x_1^2 + (x_2 + 1)^2\right)$$

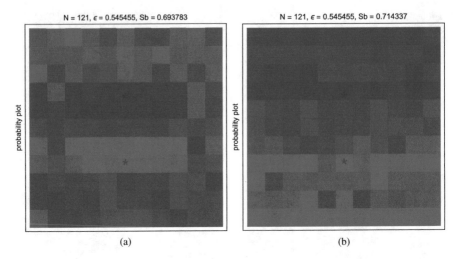

Fig. 5.8 Basin entropy of f_1 with SD$_4$ and CO. (**a**) Steepest-descent SD$_4$. (**b**) Coordinate-search CO

has a saddle point at $(0, 0.5)$ and a single global minimizer at $(0, -1)$. The saddle point and global minimizer are marked with blue and red stars, respectively, on the contour plot shown in Figure 5.9.

This function is differentiable, so we can examine its associated gradient system:

$$x_1' = -x_1 \left(4\,x_1^2 + 4\,x_2^2 + 5\right)$$
$$x_2' = \left(3 - 4\,x_1^2\right) x_2 - 4\,x_2^3 - 1 \tag{5.5}$$

whose equilibria and basins of attraction are pictured in Figure 5.10. From our discussion in Chapter 3, it should not be surprising that the minimizer in Figure 5.9 appears as an equilibrium in Figure 5.10.

5.2.1   Simulated Basins

Unlike the preceding example, we get different basin images when we apply the steepest-descent method with the exact line-search SD$_2$ and the no-longer-downhill line-search SD$_3$ to this objective function. These simulated basins of attraction are shown in Figure 5.11. Neither of these images are the same as that which we would get by simulating basins of attraction for the equilibria of the gradient system (5.5); however, the (blue) simulated basins associated with the saddle point at $(0, 0.5)$ do appear to be contained inside the corresponding basin for that equilibrium of the gradient system.

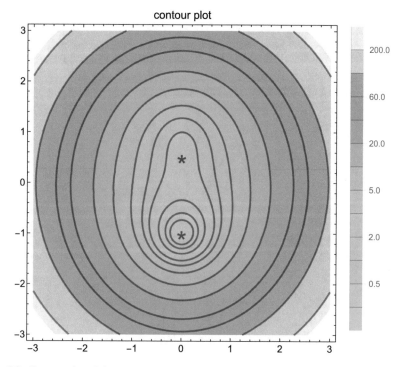

Fig. 5.9 Contour plot of f_2

If we instead apply steepest-descent with the backtracking line-search SD_4 or the coordinate-search method CS, we get the images in Figure 5.12. All of the simulated basins shown in Figures 5.11 and 5.12 are clearly different from each other, and there appears to be an empty simulated basin associated with the blue saddle point at $(0, 0.5)$ using coordinate-search.

5.2.2 Basin Sizes

Using the basin-sizing procedure described at the beginning of this section (but with 10,000 samples for SD_2 and SD_4 to get more accuracy for the very small probabilities that appeared in those cases), we get the estimates in Table 5.3 for the probabilities associated with the blue saddle point at $(0, 0.5)$ and the red global minimizer at $(0, -1)$.

We fit a line to the final four points on the $\log_2$ plot of the probability data (with $P\left(2^d\right) \neq 0$) in Table 5.3 to determine the values of γ and P_0 shown in Table 5.4. In the row corresponding to CS and the saddle point at $(0, 0.5)$, we see $P\left(2^d\right) = 0$ for all $d = 0, 1, \ldots 10$, which is consistent with the basin simulation we saw in

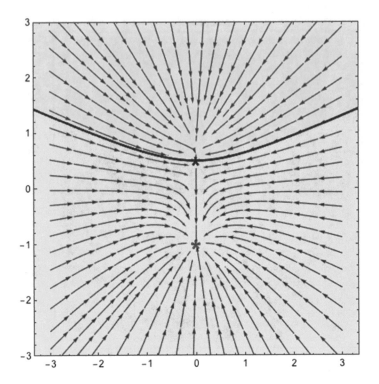

Fig. 5.10 Gradient system (5.5) associated with f_2

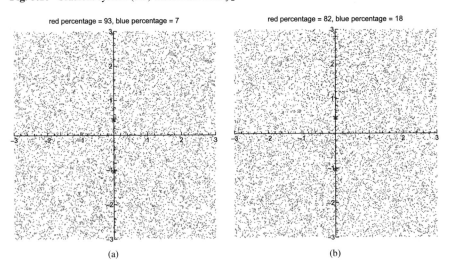

Fig. 5.11 Simulated basins of f_2 with SD_2 and SD_3. (**a**) Steepest-descent SD_2. (**b**) Steepest-descent SD_3

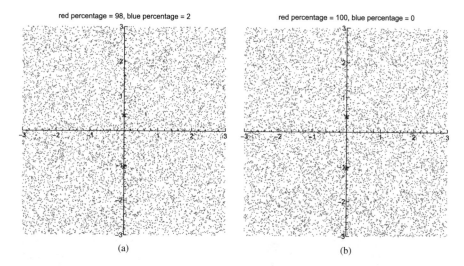

red percentage = 98, blue percentage = 2   red percentage = 100, blue percentage = 0

(a)   (b)

Fig. 5.12 Simulated basins of f_2 with SD$_4$ and CO. (**a**) Steepest-descent SD$_4$. (**b**) Coordinate-search CS

Table 5.3 Basin-size data for f_2

	d	0	1	2	3	4	5	6	7	8	9	10
SD$_2$	$P(2^d)$	0.0002	0.0003	0.0001	0.0	0.0	0.0	0.0	0.0	0.0	0.0	0.0
	$P(2^d)$	1.0	0.992	0.949	0.987	0.997	0.999	1.0	1.0	1.0	1.0	1.0
SD$_3$	$P(2^d)$	0.002	0.001	0.001	0.0	0.0	0.0	0.0	0.0	0.0	0.0	0.0
	$P(2^d)$	1.0	0.953	0.868	0.960	0.990	0.998	0.999	1.0	1.0	1.0	1.0
SD$_4$	$P(2^d)$	0.0001	0.0	0.0001	0.0	0.0	0.0	0.0	0.0	0.0	0.0	0.0
	$P(2^d)$	0.980	0.956	0.972	0.976	0.976	0.976	0.977	0.975	0.974	0.976	0.976
CS	$P(2^d)$	0.0	0.0	0.0	0.0	0.0	0.0	0.0	0.0	0.0	0.0	0.0
	$P(2^d)$	1.0	1.0	1.0	1.0	1.0	1.0	1.0	1.0	1.0	1.0	1.0

Table 5.4 Size classifications for basins of f_2

	γ	P_0	$P(r)$	Class	r_0
SD$_2$	2	0.002	$0.002\,r^{-2}$	2	0.045
	0	1.0	1.0	1	
SD$_3$	2	0.008	$0.008\,r^{-2}$	2	0.089
	0	1.0	1.0	1	
SD$_4$	2	0.001	$0.001\,r^{-2}$	2	0.032
	0	0.97	0.97	1	
CS	NA	0	0	3	
	0	1.0	1.0	1	

Figure 5.12 for the coordinate-search method. Notice that the values for the effective radius r_0 in Table 5.4 are consistent with the relative sizes of the simulated basins we see in Figures 5.11 and 5.12.

5.2.3   Basin Entropy

We computed the basin entropy associated with the square region:

$$\Omega = \{(x_1, x_2) \in \mathbb{R}^2 : -3 \leq x_1 \leq 3 \text{ and } -3 \leq x_2 \leq 3\}$$

and created the corresponding probability plots shown in Figures 5.13 and 5.14. The value of zero entropy for CO shown in Figure 5.14b is consistent with there being no basin boundaries at all in Figure 5.12.

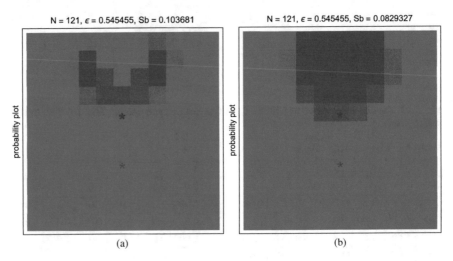

Fig. 5.13 Basin entropy of f_2 with SD$_2$ and SD$_3$. (**a**) Steepest-descent SD$_2$. (**b**) Steepest-descent SD$_3$

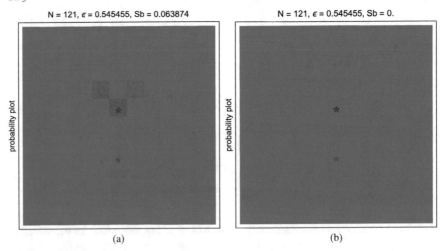

Fig. 5.14 Basin entropy of f_2 with SD$_4$ and CO. (**a**) Steepest-descent SD$_4$. (**b**) Coordinate-search CO

5.3 Example 3: One Global Minimizer and One Local Minimizer

The objective function:

$$f_3(x_1, x_2) := \frac{1}{4}\left(1 + 4\left(x_1^2 + (x_2 - 1)^2\right)\right)\left(x_1^2 + (x_2 + 1)^2\right)$$

has a global minimizer at $(0, -1)$, a local minimizer at $\left(0, \frac{1}{4}\left(2 + \sqrt{2}\right)\right)$, and a saddle point at $(0, 0)$. The local minimizer and global minimizer are marked with blue and red stars, respectively, on the contour plot shown in Figure 5.15.

This function is differentiable, so we can examine its associated gradient system:

$$x_1' = -\frac{x_1}{2}\left(8x_1^2 + 8x_2^2 + 9\right)$$

$$x_2' = \frac{1}{2}\left(\left(7 - 8x_1^2\right)x_2 - 8x_2^3 - 1\right) \tag{5.6}$$

whose equilibria and basins of attraction are pictured in Figure 5.16. From our discussion in Chapter 3, it should not be surprising that the minimizers in Figure 5.15 appear as equilibria in Figure 5.16.

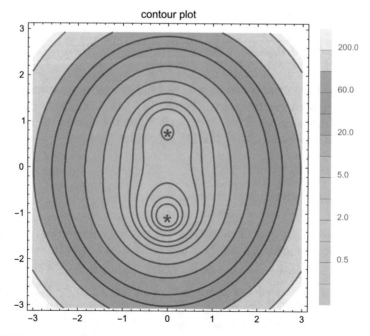

Fig. 5.15 Contour plot of f_3

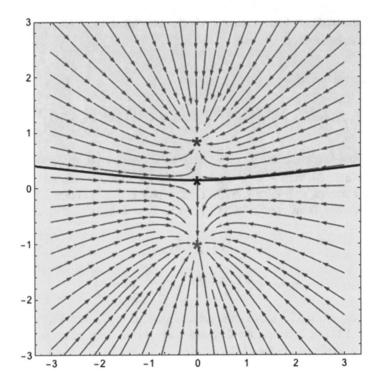

Fig. 5.16 Gradient system (5.6) associated with f_3

5.3.1 Simulated Basins

We get different simulated basins of attraction for all four of our methods for this objective function as can be seen in Figures 5.17 and 5.18. There are evidently similarities with the shapes of the corresponding simulated basins for f_2; however, the sizes of the blue basins here appear to be consistently larger.

5.3.2 Basin Sizes

Using the basin-sizing procedure described at the beginning of this section, we get the estimates in Table 5.5 for the probabilities associated with the blue local minimizer at $\left(0, \frac{1}{4}\left(2 + \sqrt{2}\right)\right)$ and the red global minimizer at $(0, -1)$.

We fit a line to the final four points (with $P\left(2^d\right) \neq 0$) on the $\log_2$ plot of the probability data in Table 5.5 to determine the values of γ and P_0 shown in Table 5.6. Note the difference in classes for the blue local minimizer at $\left(0, \frac{1}{4}\left(2 + \sqrt{2}\right)\right)$ with SD_2 and SD_3, which look in Figure 5.17 like they might belong to the same class.

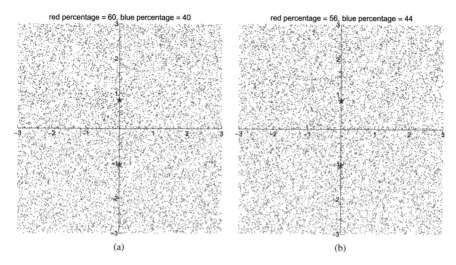

Fig. 5.17 Simulated basins of f_3 with SD$_2$ and SD$_3$. (**a**) Steepest-descent SD$_2$. (**b**) Steepest-descent SD$_3$

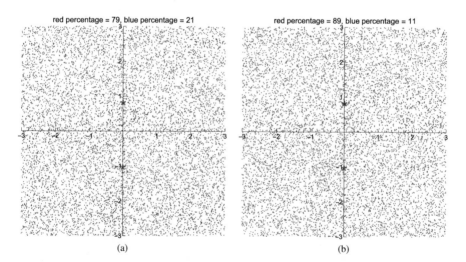

Fig. 5.18 Simulated basins of f_3 with SD$_4$ and CO. (**a**) Steepest-descent SD$_4$. (**b**) Coordinate-search CS

The wider view images in Figure 5.19 show that the first simulated basin is indeed "smaller" than the second, as the classifications in Table 5.6 suggest. In particular, the first simulated basin appears to be bounded on this scale, which is consistent with having a finite effective radius.

Note too that the larger sizes we observed for the blue basins in Figures 5.17 and 5.18 compared to Figures 5.11 and 5.12 for example function f_2 are confirmed

Table 5.5 Basin-size data for f_3

	d	0	1	2	3	4	5	6	7	8	9	10
SD_2	$P(2^d)$	0.827	0.588	0.379	0.228	0.059	0.015	0.004	0.001	0.0	0.0	0.0
	$P(2^d)$	1.0	0.848	0.775	0.810	0.938	0.984	0.996	0.999	1.0	1.0	1.0
SD_3	$P(2^d)$	0.909	0.690	0.546	0.424	0.266	0.200	0.182	0.152	0.159	0.143	0.148
	$P(2^d)$	1.0	0.858	0.730	0.708	0.773	0.810	0.825	0.840	0.819	0.838	0.853
SD_4	$P(2^d)$	0.695	0.290	0.174	0.184	0.200	0.190	0.191	0.213	0.202	0.199	0.200
	$P(2^d)$	0.890	0.735	0.749	0.799	0.792	0.793	0.797	0.802	0.800	0.792	0.798
CS	$P(2^d)$	0.135	0.120	0.131	0.127	0.110	0.102	0.112	0.113	0.128	0.115	0.117
	$P(2^d)$	1.0	0.982	0.949	0.908	0.891	0.886	0.893	0.879	0.866	0.886	0.892

Table 5.6 Size classifications for basins of f_3

	γ	P_0	$P(r)$	Class	r_0
SD_2	2	15.18	$15.18\,r^{-2}$	2	3.896
	0	1.0	1.0	1	
SD_3	0	0.18	0.18	1	
	0	0.79	0.79	1	
SD_4	0	0.24	0.24	1	
	0	0.82	0.82	1	
CS	0	0.12	0.12	1	
	0	0.83	0.83	1	

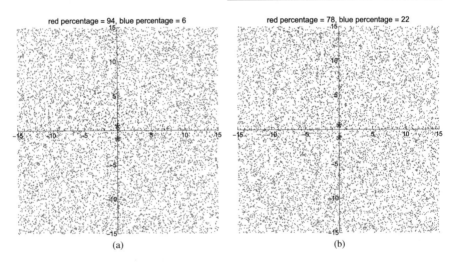

red percentage = 94, blue percentage = 6   red percentage = 78, blue percentage = 22

(a)   (b)

Fig. 5.19 Wide view of simulated basins of f_3. (**a**) Steepest-descent SD_2. (**b**) Steepest-descent SD_3

on the asymptotic scale by comparing Tables 5.4 and 5.6. For instance, there is a significantly larger effective radius for f_3 with SD_2 than for f_2.

5.3.3   Basin Entropy

We computed the basin entropy associated with the square region:

$$\Omega = \{(x_1, x_2) \in \mathbb{R}^2 : -3 \leq x_1 \leq 3 \text{ and } -3 \leq x_2 \leq 3\}$$

and created the corresponding probability plots shown in Figures 5.20 and 5.21. As in the first example, the higher entropy values for the cases shown in Figure 5.21 are consistent with the more complex basin boundaries we see in Figure 5.18.

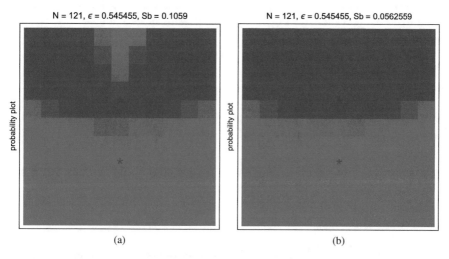

Fig. 5.20 Basin entropy of f_3 with SD$_2$ and SD$_3$. (**a**) Steepest-descent SD$_2$. (**b**) Steepest-descent SD$_3$

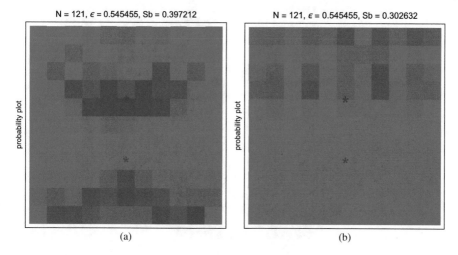

Fig. 5.21 Basin entropy of f_3 with SD$_4$ and CO. (**a**) Steepest-descent SD$_4$. (**b**) Coordinate-search CO

5.4   Example 4: One Global Minimizer and a Ring of Local Minimizers

The objective function:

$$f_4(x_1, x_2) := 4 \left(x_1^2 + x_2^2 \right)^2 + \left(x_1^2 + x_2^2 \right)^3$$

has a single global minimizer at $(0, 0)$, and a ring of local minimizers on the circle of radius $\sqrt{\frac{15+\sqrt{33}}{12}}$ about the origin. The global minimizer is marked with a red star on the contour plot shown in Figure 5.22, and the ring of local minimizers is shown in blue.

This function is differentiable, so we can examine its associated gradient system:

$$x_1' = x_1 \left(15\,x_1^2 + 15\,x_2^2 - 8 - 6 \left(x_1^2 + x_2^2 \right)^2 \right)$$

$$x_2' = x_2 \left(15\,x_1^2 + 15\,x_2^2 - 8 - 6 \left(x_1^2 + x_2^2 \right)^2 \right) \qquad (5.7)$$

whose equilibria and basins of attraction are pictured in Figure 5.23. From our discussion in Chapter 3, it should not be surprising that the minimizer in Figure 5.22 appears as equilibria in Figure 5.23.

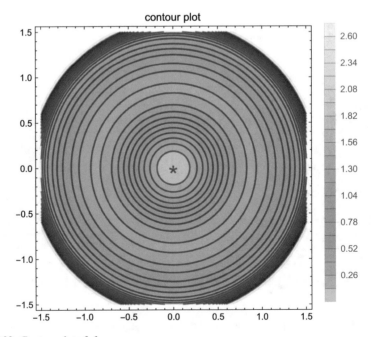

Fig. 5.22 Contour plot of f_4

5.4.1 Simulated Basins

We get different simulated basins of attraction for all four of our methods for this objective function as can be seen in Figures 5.24 and 5.25.

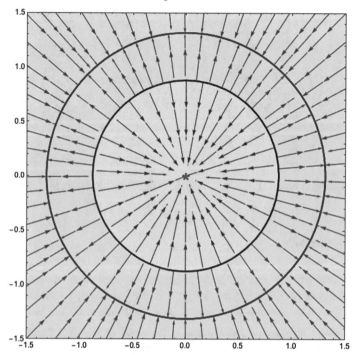

Fig. 5.23 Gradient system (5.7) associated with f_4

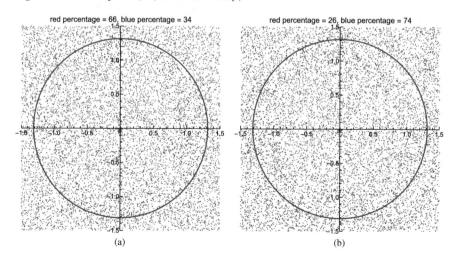

Fig. 5.24 Simulated basins of f_4 with SD_2 and SD_3. (**a**) Steepest-descent SD_2. (**b**) Steepest-descent SD_3

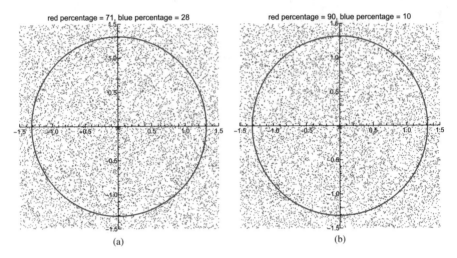

Fig. 5.25 Simulated basins of f_4 with SD$_4$ and CO. (**a**) Steepest-descent SD$_4$. (**b**) Coordinate-search CS

5.4.2 Basin Sizes

Using the basin-sizing procedure described at the beginning of this section, we get the estimates in Table 5.7 for the probabilities associated with the red global minimizer at $(0, 0)$ and the blue ring of local minimizers on the circle of radius $\sqrt{\frac{15+\sqrt{33}}{12}}$ about the origin. The change in the trend of the data for SD$_3$ suggests that we should look at more data points. If we look at ten more values of d, we get the values in Table 5.8. Wider views of the simulated basins as in Figure 5.26 show why there is a trend change in the data for SD$_3$ in this case. Note that the sampling annulus corresponding to $d = 9$ is shown here in gray since that corresponds to the trend change in the data. Note also that the effective radius of approximately 433 from Table 5.9 for this case is consistent with the radius of the "blue" region in the wide-view image of the simulated basin (Figure 5.27).

We fit a line to the final four points (with $P\left(2^d\right) \neq 0$) on the $\log_2$ plot of the probability data in Tables 5.7 and 5.8 to determine the values of γ and P_0 shown in Table 5.9. The size classifications in Table 5.9 suggest that the patterns we see in Figure 5.25 persist in wider views of the domain so that both basins are unbounded, whereas the blue simulated basin shown in Figure 5.24a is bounded and confined to the section of the domain shown there.

5.4.3 Basin Entropy

We computed the basin entropy associated with the square region:

Table 5.7 Basin-size data for f_4

	d	0	1	2	3	4	5	6	7	8	9	10
SD$_2$	$P(2^d)$	0.220	0.242	0.060	0.015	0.004	0.001	0.0	0.0	0.0	0.0	0.0
	$P(2^d)$	0.768	0.762	0.940	0.985	0.996	0.999	1.0	1.0	1.0	1.0	1.0
SD$_3$	$P(2^d)$	0.021	0.802	0.951	0.988	0.997	0.999	1.0	1.0	1.0	0.711	0.179
	$P(2^d)$	0.785	0.196	0.049	0.012	0.003	0.001	0.0	0.0	0.0	0.268	0.817
SD$_4$	$P(2^d)$	0.127	0.239	0.206	0.192	0.207	0.201	0.195	0.199	0.216	0.191	0.190
	$P(2^d)$	0.814	0.755	0.791	0.783	0.799	0.788	0.791	0.805	0.807	0.787	0.800
CS	$P(2^d)$	0.0	0.110	0.171	0.209	0.205	0.220	0.225	0.234	0.234	0.246	0.235
	$P(2^d)$	1.0	0.878	0.831	0.796	0.774	0.792	0.767	0.779	0.765	0.773	0.771

Table 5.8 Extra basin size data for f_4 with SD$_3$

	d	10	11	12	13	14	15	16	17	18	19	20
SD$_3$	$P(2^d)$	0.045	0.011	0.003	0.001	0.0	0.0	0.0	0.0	0.0	0.0	0.0
	$P(2^d)$	0.954	0.989	0.997	0.999	1.0	1.0	1.0	1.0	1.0	1.0	1.0

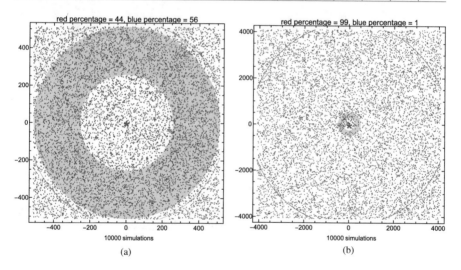

Fig. 5.26 Wide-view simulated basins of f_4 with SD$_3$. (**a**) Wide view. (**b**) Wider view

$$\Omega = \{(x_1, x_2) \in \mathbb{R}^2 : -1.5 \le x_1 \le 1.5 \text{ and } -1.5 \le x_2 \le 1.5\}$$

and created the corresponding probability plots shown in Figures 5.27 and 5.28. In the case of SD$_4$ shown in Figure 5.28a, we see an entropy value that is relatively close to 1, which is the highest entropy possible. This is consistent with the complex basin boundary we see for this case in Figure 5.25a.

Table 5.9 Size
classifications for basins of f_4

	γ	P_0	$P(r)$	Class	r_0
SD_2	2	0.97	$0.97\,r^{-2}$	2	0.983
	0	1.0	1.0	1	
SD_3	2	187,236	$187,236\,r^{-2}$	2	432.707
	0	1.0	1.0	1	
SD_4	0	0.25	0.25	1	
	0	0.83	0.83	1	
CS	0	0.23	0.23	1	
	0	0.79	0.79	1	

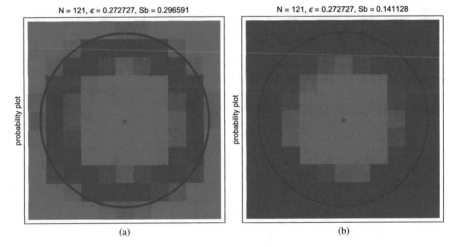

Fig. 5.27 Basin entropy of f_4 with SD_2 and SD_3. (**a**) Steepest-descent SD_2. (**b**) Steepest-descent SD_3

5.5   Nelder–Mead Method

In this section, we apply the Nelder–Mead method NM to all four of the preceding
example objective functions and investigate the resulting basins of attraction in each
case.

5.5.1   Simulated Basins

Since the iterate-multisets in this case consist of $m = 3$ elements in $\mathbb{R}^2$, the basins
of attraction essentially inhabit $\mathbb{R}^6$. We cannot visualize this space, but there are a
variety of ways to attempt to illustrate simulated basins of attraction. One possibility
is with three copies of $\mathbb{R}^2$ stacked together as in Figure 5.29. The blue-shaded
triangles on top represent the initial iterate-multisets attracted to the blue minimizer

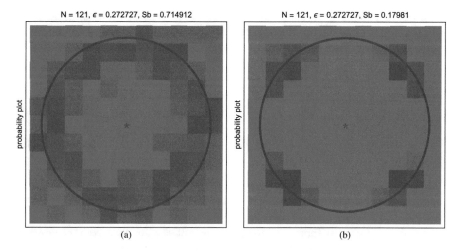

N = 121, ϵ = 0.272727, Sb = 0.714912   N = 121, ϵ = 0.272727, Sb = 0.17981

Fig. 5.28 Basin entropy of f_4 with SD_4 and CO. (**a**) Steepest-descent SD_4. (**b**) Coordinate-search CO

Fig. 5.29 Simulated basins of f_1 with NM

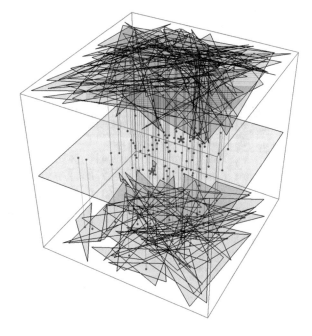

at $(0, 1)$, and the red-shaded triangles on the bottom represent the initial iterate-multisets attracted to the red minimizer at $(0, -1)$. The copy of the x_1x_2-plane in the middle shows the two minimizers as well as blue and red points at the centers of each of the blue-shaded and red-shaded triangles above and below. While this illustration is an accurate representation of the objects contained in the simulated basins of attraction, it can be difficult to observe any meaningful patterns here. We will develop a more useful illustration of the simulated basins of attraction in the forthcoming section on basin entropy.

5.5.2 Basin Sizes

Using the basin-sizing procedure described at the beginning of this section with the excess pre-distance function:

$$\mathbf{d}_{\text{exc}}(X^0, \bar{x}^\infty) = \sup_{x \in X^0} \|x - \bar{x}^\infty\|,$$

we get the estimates in Table 5.10 for the probabilities associated with the attractor sets associated with the four preceding examples in this section.

We fit a line to the final four points (with $P_{\mathbf{d}_{\text{exc}}}\left(2^d\right) \neq 0$) on the $\log_2$ plot of the probability data in Table 5.10 to determine the values of γ and P_0 shown in Table 5.11.

Table 5.10 Basin-size data for example objectives with NM

	d	0	1	2	3	4	5	6	7	8	9	10
f_1	$P_{\mathbf{d}_{\text{exc}}}\left(2^d\right)$	0.894	0.702	0.571	0.528	0.513	0.503	0.490	0.461	0.488	0.500	0.501
	$P_{\mathbf{d}_{\text{exc}}}\left(2^d\right)$	0.907	0.745	0.597	0.545	0.511	0.507	0.518	0.498	0.505	0.503	0.505
f_2	$P_{\mathbf{d}_{\text{exc}}}\left(2^d\right)$	0.0	0.0	0.0	0.0	0.0	0.0	0.0	0.0	0.0	0.0	0.0
	$P_{\mathbf{d}_{\text{exc}}}\left(2^d\right)$	1.0	1.0	1.0	1.0	1.0	1.0	1.0	1.0	1.0	1.0	1.0
f_3	$P_{\mathbf{d}_{\text{exc}}}\left(2^d\right)$	0.721	0.470	0.294	0.209	0.159	0.149	0.146	0.157	0.165	0.149	0.136
	$P_{\mathbf{d}_{\text{exc}}}\left(2^d\right)$	0.984	0.901	0.833	0.827	0.827	0.839	0.856	0.847	0.859	0.862	0.847
f_4	$P_{\mathbf{d}_{\text{exc}}}\left(2^d\right)$	0.048	0.235	0.277	0.183	0.133	0.106	0.096	0.090	0.081	0.089	0.086
	$P_{\mathbf{d}_{\text{exc}}}\left(2^d\right)$	0.955	0.766	0.744	0.817	0.874	0.905	0.907	0.912	0.931	0.930	0.924

Table 5.11 Size classifications for basins of example objectives with NM

	γ	P_0	$P_{\mathbf{d}_{\text{exc}}}(r)$	Class	r_0
f_1	0	0.39	0.39	1	
	0	0.48	0.48	1	
f_2	NA	0	0	3	
	0	1.0	1.0	1	
f_3	0.1	0.24	$0.24\,r^{-0.1}$	2	0.0
	0	0.85	0.85	1	
f_4	0	0.09	0.09	1	
	0	0.90	0.90	1	

5.5.3 Basin Entropy

As we have seen in the preceding sections on basin entropy, probability plots can provide a reasonable (albeit crude) illustration of simulated basins of attraction, since basin boundaries tend to be the only high-entropy regions. In this section, we will develop a nested version of probability plots that provides a useful illustration of simulated basins of attraction even when the iterate-multisets have cardinality $m > 1$.

Since the Nelder–Mead method applied to any of our four example objectives uses $m = 3$ elements in each iterate-multiset, we subdivide $\Omega^3 = \Omega \times \Omega \times \Omega \subseteq \mathbb{R}^6$ into a grid of equal hypercubes in $\mathbb{R}^6$ with edge-length ϵ. We can consider each hypercube B as the cross-product of three squares S_i in $\mathbb{R}^2$: $B = S_1 \times S_2 \times S_3$; each with edge-length ϵ. Thus, when we take a pseudo-random sample:

$$X^0 = (x_1^0, x_2^0, x_3^0, x_4^0, x_5^0, x_6^0)$$

from B to generate our initial iterate-multiset of three vertices in $\mathbb{R}^2$ for NM, we are essentially sampling pseudo-randomly from S_1 to get our first vertex $v_1^0 := (x_1^0, x_2^0)$, sampling pseudo-randomly from S_2 to get our first vertex $v_2^0 := (x_3^0, x_4^0)$, and sampling pseudo-randomly from S_3 to get our first vertex $v_3^0 := (x_5^0, x_6^0)$.

In order to visualize our results, we use a schematic overlaying three layers of different-scale Ω. On the largest scale is the copy of Ω containing a grid of squares S_1 from which the first vertices v_1^0 are generated. On the next scale, we fit an entire shrunken copy of the analogous grid of squares S_2 exactly into each square S_1. On the smallest scale, we then fit an entire shrunken copy of the grid of squares S_3 exactly into each square S_2. The location of each of these smallest squares in the schematic then corresponds to the location of a set of three vertices: v_1^0 coming from the square S_1 on the largest scale, v_2^0 coming from the square S_2 located in the position relative to subgrid on the middle level, and v_3^0 coming from the square S_3 located in the position relative to subgrid on the lowest level.

For the first three example objective functions, we use $\epsilon = 2$ and $\Omega = [-3, 3] \times [-3, 3]$, so there are nine total squares S_1 on the largest scale. Each of these S_1 contains nine smaller versions of S_2, and each such S_2 contains nine even smaller versions of S_3. For the fourth example objective function, we use $\epsilon = 1$ and $\Omega = [-1.5, 1.5] \times [-1.5, 1.5]$ for the same breakdown of nine squares in each subgrid. In all four cases, this results in $9^3 = 729$ total hypercubes in $\mathbb{R}^6$, which corresponds to 729 squares on the smallest scale of our schematic. We take 25 pseudo-randomly generated samples from each hypercube, and we color-code each of the smallest squares on our schematic according to the fraction of the corresponding initial iterate-multisets that are attracted to the red or blue attractor sets. Our results are displayed in Figures 5.30 and 5.31. Notice that the grid lines for the sub-levels are shown in order to make it easier to read the schematic. For instance, the mostly red square on the smallest scale six squares from the left on the bottom row in Figure 5.30a, which indicates that most of the pseudo-randomly

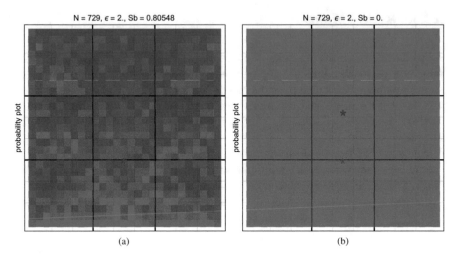

Fig. 5.30 Basin entropy of f_1 and f_2 with NM. (**a**) Example objective f_1. (**b**) Example objective f_2

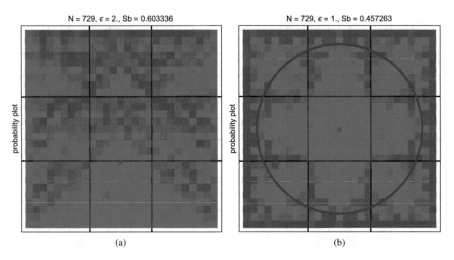

Fig. 5.31 Basin entropy of f_3 and f_4 with NM. (**a**) Example objective f_3. (**b**) Example objective f_4

generated initial iterate-multisets were attracted to the red minimizer at $(0, -1)$ when their first vertex came from the lower-left square S_1 of the grid, their second vertex came from the lower-middle square S_2 of the grid, and their third vertex came from the lower-right square S_3 of the grid.

The all-red squares and entropy of $Sb = 0$ in Figure 5.30b suggest that all pseudo-randomly generated initial iterate-multisets are attracted to the minimizer at $(0, -1)$ for f_2 with the Nelder–Mead method (as was the case with coordinate-search CO). The mostly blue squares around the perimeter in Figure 5.31b suggest that most pseudo-randomly generated initial iterate-multisets are attracted to the ring of local minimizers for f_4 when their vertices all come from the perimeter of the grid.

5.6 Practical Significance of Counterexamples

In this section, we will investigate several counterexamples to conjectures about what conditions on objective functions might ensure that initial iterate-multisets generated by particular minimization methods are attracted to minimizers. The counterexamples invalidate the conjectures regardless of how many initial iterate-multisets fail to be attracted to a minimizer. However, we can use the tools we have developed for investigating basins of attraction to refine our understanding of the "practical significance" of each counterexample, where a practically significant counterexample is one for which the undesirable behavior is likely to occur should someone attempt to apply the method to the objective function.

Both of the counterexamples we investigate here have a single minimizer x^∞, and we use its basin of attraction $\text{Basin}_{\mathcal{M},f}(x^\infty)$ to measure practical significance both on the asymptotic scale and on the scale of the bounded region:

$$\Omega = \{(x_1, x_2) \in \mathbb{R}^2 : -3 \le x_1 \le 3 \text{ and } -3 \le x_2 \le 3\}$$

of the domain space $\mathbb{R}^n$. For counterexamples with more minimizers (hence more "desirable" outcomes), we would expand our analysis to the basins associated with those as well.

On the asymptotic scale, we simply determine the basin size of $\text{Basin}_{\mathcal{M},f}(x^\infty)$ according to our classifications, and note that a smaller basin size then signals greater practical significance of the counterexample on this scale. On the scale of Ω, we calculate the ratio of basin entropy Sb to the percentage of pseudo-random initial iterate-multisets in Ω that are in $\text{Basin}_{\mathcal{M},f}(x^\infty)$. A large entropy (numerator) combined with a small percentage (denominator) corresponds to a relatively diffuse and sparse $\text{Basin}_{\mathcal{M},f}(x^\infty)$ on this scale. Thus, the larger this ratio is, the greater the practical significance of the counterexample on the scale of Ω.

5.6.1 Coordinate-Search and the Canoe Function

The function:

$$f(x_1, x_2) := \frac{1}{2} \max\left\{(x_1 - 1)^2 + (x_2 + 1)^2, (x_1 + 1)^2 + (x_2 - 1)^2\right\}$$

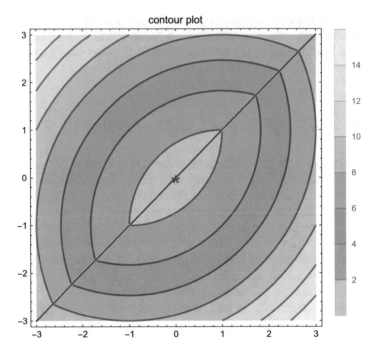

Fig. 5.32 Contour plot of canoe function

is a variant from [14] of the Dennis–Woods function [10], and it is called the "canoe" function because of the shape of its contours (e.g., see Figure 5.32). Even though the canoe function is continuous and strictly convex, the coordinate-search method can fail to identify its global minimizer at $(0, 0)$. Initial iterates are attracted to the line $x_2 = x_1$ (colored blue in Figure 5.32), but not necessarily to the global minimizer at $(0, 0)$ (the red star in Figure 5.32). This example establishes a limitation on what convergence results may be proven for coordinate-search. In particular, the canoe function is a counterexample to the conjecture that continuity and strict convexity of the objective function are enough to ensure that the iterates generated by the coordinate-search method will be attracted to a minimizer. We will now see that this counterexample is practically significant, in part because most initial iterates are not attracted to the red minimizer $(0, 0)$ and instead are attracted to the blue attractor set:

$$S := \left\{ (x_1, x_2) \in \mathbb{R}^2 : x_2 = x_1 \right\} \setminus \{(0, 0)\}. \tag{5.8}$$

This is clear on the scale shown in Figure 5.33, and persists on larger scales as we can see from the basin-size analysis shown in Tables 5.12 and 5.13. Notice that the asymptotic models $P(r)$ for the probabilities associated with each attractor set determined by the values in Table 5.13 suggest that it is much more likely (95% versus 3%) for an initial iterate to be attracted to the blue attractor set S (5.8) than for

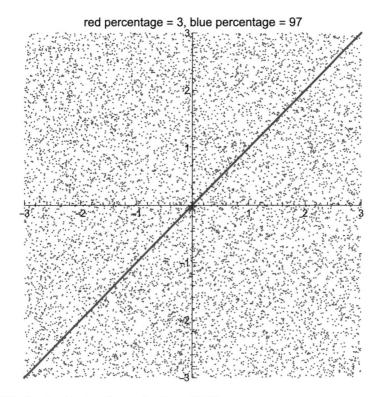

Fig. 5.33 Simulated basins of canoe function with CO

Table 5.12 Basin-size data for canoe function with CO

d	0	1	2	3	4	5	6	7	8	9	10
$P(2^d)$	0.975	0.971	0.969	0.968	0.963	0.958	0.972	0.968	0.961	0.970	0.973
$P(2^d)$	0.039	0.041	0.032	0.034	0.040	0.033	0.034	0.032	0.036	0.032	0.034

Table 5.13 Size classifications for basins of canoe function with CO

γ	P_0	$P(r)$	Class	r_0
0	0.95	0.95	1	
0	0.03	0.03	1	

it to be attracted to the red minimizer at $(0, 0)$. Notice too that these percentages are consistent in this case with the percentages seen on the smaller scale in Figure 5.33.

The probability plot shown in Figure 5.34 suggests that initial iterates attracted to the red minimizer at $(0, 0)$ are relatively diffuse and sparse on this scale. The basin entropy value of $Sb = 0.187479$ in this case, divided by the red percentage of 0.03 from Figure 5.33 gives a relatively large ratio of 6.2493, which is consistent with $\text{Basin}_{\mathcal{M},f}(0, 0)$ being relatively diffuse and sparse on this scale. All of our evidence together suggests that this counterexample is practically significant, since the undesirable behavior (of the pseudo-randomly initiated method identifying a non-minimizer) is likely to occur.

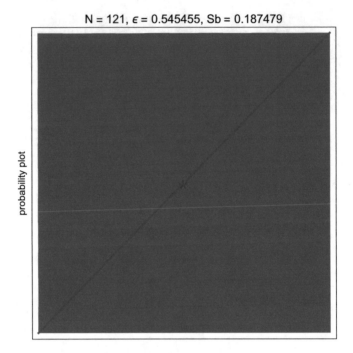

Fig. 5.34 Basin entropy of canoe function with CO

Table 5.14 Basin-size data for McKinnon's function ($\tau = 2$, $\theta = 6$, and $\phi = 60$) with NM

d	0	1	2	3	4	5	6	7	8	9	10
$P_{d_{exc}}\left(2^d\right)$	0.0	0.0	0.0	0.0	0.0	0.0	0.0	0.0	0.0	0.0	0.0
$P_{d_{exc}}\left(2^d\right)$	1.0	1.0	1.0	1.0	1.0	1.0	1.0	1.0	1.0	1.0	1.0

5.6.2   The Nelder–Mead Method and McKinnon's Function

We have seen in Section 4.5 that the Nelder–Mead method applied to McKinnon's [17] function (with $\tau = 2$, $\theta = 6$, and $\phi = 60$) has an attractor set (in blue on the contour plot in Figure 4.1) as well as a red attractor at the minimizer $(0, -0.5)$. The pseudo-random initial iterate-multisets from all 1000 simulations were in the basin of attraction $\text{Basin}_{\mathcal{M}, f}(0, -0.5)$ associated with the minimizer $(0, -0.5)$. In particular then, we get the basin-size data in Table 5.14. As a result, we see in Table 5.15 that $\text{Basin}_{\mathcal{M}, f}(0, -0.5)$ is as large as possible, which means that this counterexample is not practically significant on the asymptotic scale. The probability plot in Figure 5.35 suggests that the basin $\text{Basin}_{\mathcal{M}, f}(0, -0.5)$ is neither diffuse nor sparse on this scale. The entropy of $Sb = 0$ and the red percentage of 1.0 have a ratio of 0, which corresponds to the smallest possible practical significance on this scale.

Table 5.15 Size classifications for basins of McKinnon's function ($\tau = 1$, $\theta = 6$, and $\phi = 50$) with NM

γ	P_0	$P_{d_{exc}}(r)$	Class	r_0
NA	0.0	0.0	3	
0	1.0	1.0	1	

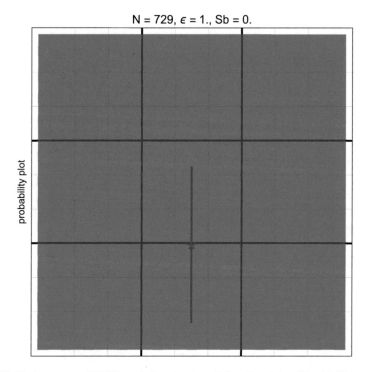

Fig. 5.35 Basin entropy of McKinnon's function ($\tau = 2$, $\theta = 6$, and $\phi = 60$) with NM

Thus, we see that this counterexample is not practically significant on either the asymptotic or the bounded scale. This means that the undesirable behavior (of the pseudo-randomly initiated method identifying a non-minimizer) is very unlikely to occur, suggesting that McKinnon's counterexample does not cast doubt in the reliability of the Nelder–Mead method in practice, and supporting the continued popularity of the Nelder–Mead method.

Our investigation of this example was stimulated by the curious fact that the Nelder–Mead method is widely used despite its possibility of failure for high-dimensional problems [23], as well as the impossibility of traditional convergence theory to support it (as demonstrated by McKinnon [17]). The authors of [15] suggest some possible reasons for the continued popularity of NM, and we have just added to their list by demonstrating how the theory-stifling result of [17] can be viewed as insignificant in practice where the essential issue is the extent to which initial data are likely to lead to undesirable consequences.

References

1. Abramson, M.A.: Second-order behavior of pattern search. SIAM J. Optim. **16**, 315–330 (2005)
2. Absil, P.-A., Kurdyka, K.: On the stable equilibrium points of gradient systems. Syst. Control Lett. **55**, 573–577 (2006)
3. Asenjo, D., Stevenson, J., Wales, D., Frenkel, D.: Visualizing basins of attraction for different minimization algorithms. J. Phys. Chem. B **117**, 12717–12723 (2013)
4. Aubin, J.-P., Frankowska, H.: Set-Valued Analysis. Birkhäuser Boston, Inc., Boston, MA (1990)
5. Banach, S.: Sur les opérations dans les ensembles abstraits et leur application aux équations intégrales. Fundam. Math. **3**, 133–181 (1922)
6. Berge, C.: Espaces Topologiques: Fonctions Multivoques. Dunod, Paris (1959)
7. Blanchard, P., Devaney, R.L., Hall, G.R.: Differential Equations, 4th edn. Brooks/Cole, Cengage Learning, Boston (2012)
8. Ciarlet, P.G., Raviart, P.-A.: General Lagrange and Hermite interpolation in $\mathbb{R}^n$ with applications to finite element methods. Arch. Ration. Mech. Anal. **46**, 177–199 (1972)
9. Daza, A., Wagemakers, A., Georgeot, B., Guéry-Odelin, D., Sanjuán, M.A.F.: Basin entropy: a new tool to analyze uncertainty in dynamical systems. Sci. Rep. **6**, 31416 (2016)
10. Dennis Jr., J.E., Woods, D.J.: Optimization on microcomputers: the Nelder-Mead simplex algorithm. In: Wouk, A. (ed.) New Computing Environments: Microcomputers in Large-Scale Computing. SIAM, Philadelphia (1987)
11. Hooke, R., Jeeves, T.A.: "Direct search" solution of numerical and statistical problems. J. Assoc. Comput. Mach. **8**, 212–229 (1961)
12. Klein, E., Thompson, A.C.: Theory of Correspondences. Including Applications to Mathematical Economics. Wiley, New York (1984)
13. Knuth, D.E.: Seminumerical Algorithms. The Art of Computer Programming, vol. 2, 3rd edn. Addison-Wesley, Boston (1998)
14. Kolda, T.G., Lewis, R.M., Torczon, V.: Optimization by direct search: new perspectives on some classical and modern methods. SIAM Rev. **45**, 385–348 (2003)
15. Lagarias, J.C., Reeds, J.A., Wright, M.H., Wright, P.E.: Convergence properties of the Nelder-Mead simplex method in low dimensions. SIAM J. Optim. **9**, 112–147 (1998)
16. Levy, A.B.: Stationarity and Convergence in Reduce-or-Retreat Minimization. Springer Briefs in Optimization. Springer, New York (2012)
17. McKinnon, K.I.M.: Convergence of the Nelder-Mead simplex method to a nonstationary point. SIAM J. Optim. **9**, 148–158 (1998)

18. Mordukhovich, B.S.: Variational Analysis and Generalized Differentiation. I. Basic Theory & II. Applications. Springer, Berlin (2006)
19. Nelder, J.A., Mead, R.: A simplex method for function minimization. Comput. J. **7**, 308–313 (1965)
20. Robinson, S.M.: Generalized equations and their solutions. I. Basic theory. In: Point-to-Set Maps and Mathematical Programming. Mathematical Programming Studies, vol. 10, pp. 128–141. Springer, Berlin (1979)
21. Rockafellar, R.T., Wets, R.J.-B.: Variational Analysis. Springer, Berlin (1998)
22. Sprott, J.C., Xiong, A.: Classifying and quantifying basins of attraction. Chaos **25**, 083101 (2015)
23. Torczon, V.: Multi-directional search: a direct search algorithm for parallel machines. Ph.D. thesis, Rice University, Houston, TX (1989)
24. Vasilesco, F.: Essai sur les fonctions multiformes de variables réelles. Thèse, Université de Paris. Gauthier-Villars, Paris (1925)
25. Zangwill, W.I.: Convergence conditions for nonlinear programming algorithms. Manag. Sci. **16**, 1–13 (1969)

Printed in the United States
By Bookmasters